Topics in Fluid Mechanics

Topics in
Fluid Mechanics

RENE CHEVRAY
COLUMBIA UNIVERSITY

JEAN MATHIEU
ÉCOLE CENTRALE DE LYON

CAMBRIDGE UNIVERSITY PRESS
Cambridge, New York, Melbourne, Madrid, Cape Town,
Singapore, São Paulo, Delhi, Tokyo, Mexico City

Cambridge University Press
The Edinburgh Building, Cambridge CB2 8RU, UK

Published in the United States of America by
Cambridge University Press, New York

www.cambridge.org
Information on this title: www.cambridge.org/9780521422727

© Cambridge University Press 1993

First published 1993

A catalogue record for this publication is available from the British Library

Library of Congress Cataloguing in Publication data
Chevray, Rene.
  Topics in fluid mechanics / Rene Chevray, Jean Mathieu.
    p. cm.
  Includes index.
  ISBN 0-521-41082-7 (hc). – ISBN 0-521-42272-8 (pb)
    1. Fluid mechanics.   I. Mathieu, Jean, 1924–   .   II. Title.
QA901.C525   1993
532–dc20   92-10886
                                              CIP

ISBN 978-0-521-41082-3 Hardback
ISBN 978-0-521-42272-7 Paperback

Contents

Preface   *page* ix

Notation   xiii

1 Basic Concepts   **1**

   1.1 Continuum   1
   1.2 Description of fluid motion   1
   1.3 Pathlines, streaklines, and streamlines   3
   1.4 Example of flow lines   4
   1.5 Visualization of streamlines, pathlines, and streaklines   6
   1.6 Basic equations for an inviscid fluid   7

2 Toward a Constitutive Law for Fluid Materials   **13**

   2.1 Introduction   13
   2.2 Volume forces and surface forces acting on a fluid   13
   2.3 Representation of surface forces by the stress tensor   14
   2.4 The equations of motion   19
   2.5 Preliminary remarks on the internal properties of
       continuous media   21
   2.6 The Navier–Stokes equations   25
   2.7 Generalized Newtonian fluids   31
   2.8 Memory effects   34
   2.9 Influence of scaling on the description of a
       physical phenomenon   40

3 Vorticity Laws   **45**

   3.1 Introduction   45
   3.2 Crocco's theorem   46
   3.3 Basic equations   47
   3.4 The vorticity flux across a closed surface is null   51
   3.5 Three theorems on the development of vortex lines   54
   3.6 The velocity field deduced from the vorticity field   61
   3.7 Implications of the vortex theory in complex
       computations   66

4   Fundamental Concepts of Turbulence   **77**

4.1   Introduction   77
4.2   A statistical approach to turbulent flow   81
4.3   The scaling of turbulence   90
4.4   Global behavior of turbulent boundary layers   101
4.5   The concept of boundary layer in turbulent flows:
        Case of the free jet   112

5   Dynamical Systems and Chaotic Advection   **121**

5.1   Introduction   121
5.2   Dynamical systems   122
5.3   Quantifying chaos   126
5.4   Strange attractors and the Lorenz system   131
5.5   Chaos in fluids   133

6   Chaos and the Onset of Turbulence   **141**

6.1   Hamiltorian systems and dissipative systems   141
6.2   Routes to turbulence   142
6.3   Routes to chaos   144
6.4   Chaos quantification   149
6.5   Application to fluid mechanics   155

7   Boundary Layer Theory   **163**

7.1   Some general remarks   163
7.2   Introduction to development of flow with vorticity   164
7.3   Boundary layer approximation   167
7.4   The mathematical structure of the equations   170
7.5   An introduction to singular perturbation   175
7.6   The boundary layer equations   180
7.7   Coupled phenomena   192
7.8   Combustion problems in fluid mechanics   198
7.9   Three-dimensional boundary layers   207

8   Wave Propagation   **211**

8.1   Introduction   211
8.2   Sound waves   212
8.3   Surface disturbances and wave speed   215
8.4   Internal waves   218
8.5   Pressure wave in an elastic duct   224
8.6   Alfvén waves   226
8.7   Wave scattering due to turbulence   233
8.8   Waves of finite amplitude   238

9 Fundamentals of Classical Thermodynamics   **241**

    9.1   Introduction   241
    9.2   The first law of thermodynamics   246
    9.3   The second law of thermodynamics   253
    9.4   The energy equation at a point   263

Appendixes   **273**

    A   The concept of "objective time derivatives"   275
    B   Partial differential equations of mathematical physics   281
    C   Perturbation theory   301

Index   319

Preface

Although many fluid mechanics textbooks exist at the undergraduate level, there are few graduate level texts. Most of these few treat the subject in a very ordinary way. Our aim in this text is not to offer a classical view of an established subject for the purpose of solving well-posed conventional exercises, but to present several subjects chosen from among currently important topics in fluid mechanics. A unity of concepts, rather than of material, connects the various subjects that are discussed. Our emphasis on concept may make it difficult to read an isolated chapter. For instance, the discussion of boundary layers that exhibit fading memory is not comprehensible without having first read the chapter on rheological laws.

Knowledge of elementary concepts is assumed, and basic notions acquired at the undergraduate level are expanded to introduce new concepts that are useful for tackling problems in basic research and industry. The Navier–Stokes equations, for example, which are usually introduced for materials that exhibit no memory effects are here extended to include them, leading to higher-order equations. Similarly, the interplay between convective and diffusive effects in many flows is introduced. Mathematical developments are kept to a minimum by placing details of important concepts in appendixes. Linear operators such as the divergence and curl should prepare the reader for concepts such as the advective operator, which becomes nonlinear when applied to velocity (or vorticity) fields. With such tools, coupling mechanisms between advective and diffusive processes are more easily understood.

Fundamental ideas are reviewed in the first two chapters. Rheological laws are stated. When a detailed description of the flow is not needed, we observe that all motions that do not alter the relative positions of marked particles do not modify the constitutive laws. In this context, we introduce the principle of objectivity.

In Chapter 3 on vorticity, we present the role of pressure, which is linked to the whole velocity field through a Poisson equation, and we explain the kinematic properties of the field by means of the curl operator. The vorticity laws introduced are those proposed by Lagrange, Helmholtz, and Cauchy. Among them, Cauchy's approach, which refers to a lagrangian description of fluid motion, is by far the most sophisticated. In all of those laws, however, it is the interaction between the

ix

velocity field and the vorticity fields, introduced as a consequence of the nonlinearity of the fundamental equations, that we present in detail.

Turbulence is presented in Chapter 4, in which we introduce scaling laws. For newtonian fluids, no memory effects appear in the rheological law; but when we introduce statistical moments, memory effects do occur, and the statistical moments are governed by partial differential equations. There is no simple relation connecting fluctuating motion to mean velocity gradients in this case, because the characteristic times due to the turbulent velocity field and the second-order moments are of the same order of magnitude. Consequently, the double velocity correlations are not instantaneously related to the mean field. We emphasize here the concept of characteristic times, which play an important role. In the flow-field description, we show that an advective mechanism associated with small scales competes with a diffusion process that is controlled by kinematic viscosity. This sort of competition appears in many flows and is described in detail in Chapter 7, on the boundary layer. Similar behavior is observed in turbulent flows for scalar quantities, and the spectral laws proposed by Corrsin and Batchelor are also presented.

A relatively new concept–that of chaotic advection–is introduced in Chapter 5. That such a presentation is made in this otherwise standard text is because it is rare that such a novel and simple concept can find direct applications to a large and important class of problems related especially to mixing. Chapter 6, on chaos and the onset of turbulence, is based on fundamental concepts of dynamical system theory presented in the preceding chapter. This approach to a large class of fluid mechanics problems is still in its infancy, but it deserves mention in a fluid mechanics text. Such a presentation could have been included in the chapter on turbulence; but because the relation between the two subjects is still a matter of conjecture, we feel that this discussion should stand on its own at present.

In Chapter 7, on boundary layers, we emphasize the interplay between the advection phenomenon and the associated diffusion terms in the region close to the wall. Here again we introduce inner and outer operators, as well as the concept of scaling associated with the thickness of the boundary layer. The matching process and a brief introduction to partial differential equations are presented in two appendixes. Flames are introduced here because, though advective effects dominate upstream and downstream of the flame front, for a flame the advection operator must be supplemented by a diffusion operator as in the boundary layer. Moreover, this flow is interesting because, in addition to the aforementioned processes, the chemical reaction introduces a characteristic time. These three mechanisms lead to relationships for the flame thickness, speed, and so on.

With the same idea in mind, of offering a unified treatment of a whole field, we present sound waves, surface and internal waves, and Alfén waves in Chapter 8, on wave propagation. The role of a wave operator is here associated with the process of energy transfer. As the medium becomes more complex, the feasibility of energy storage appears; we present Alfén waves to illustrate this fact. Interactions between waves and random media are briefly discussed, and the interplay between turbulence and an ordered mechanism is presented. Readers who are familiar with the concept of response function will find examples here.

Finally, we present a synthetic view of classical thermodynamics in Chapter 9, beginning with the first and second principles. That these two principles are complementary is demonstrated through classical examples. Even when the amount of energy is kept constant, its "quality" is eroded by the combined action of irreversible exchanges and frictional forces. Here, it is the energy equation at a particular point that is important. The choice of a constitutive law is not as free as first expected; in fact, it is limited to the second principle. This feature appears in all situations where countergradient transport occurs. We briefly touch on the compatibility of constitutive laws, including fading memories; however, it is the global problem that is important.

This text is not oriented toward particular applications but provides a manner of thinking by developing a few fundamental concepts and applying them to typical cases. The operator concept is helpful, but we consider its physical meaning to be more important than its mathematical treatment. We use the operator extensively because it is not linked to special features and because it is capable of embodying large classes of phenomena. General concepts help relate all parts of this course. This point of view, we feel, is more insightful than the usual one.

This text is an amalgamation of courses being taught by one of us (JM) at the École Centrale de Lyon for the last twenty years and by the other (RC) at the Department of Mechanical Engineering at Columbia University for the last ten years. The idea of combining the two courses into a text stemmed from similar points of view in presentation. The mechanism of writing this text was begun while JM was a visiting professor at Columbia in the spring of 1987.

We thank the many people whose comments about the unpublished version of the book have been so helpful. Among them, we thank especially Pradip Dutta and Cyril Volte who read one of the latest versions of the text. Special thanks are due to long-time colleagues and friends: Professor S.-I. Iida, who has always shown great interest and enthusiam for chaotic flows, Professor P. Bourgin, who contributed instructive discussions on the behavior of materials, and Dr. J.-C. Tatinclaux, who has made many corrections and suggestions for the last version of the text. This book has greatly benefited from our professional association with them.

Finally, it has been our pleasure to work with a very understanding staff at Cambridge University Press. In particular, our editor Florence Padgett showed patience and understanding during all stages necessary to bring the book into production.

New York   R. C. and J. M.
December 18, 1991

Notation

a	speed of sound
A	area, arbitrary constant
B	arbitrary constant
c	wave phase speed
C	concentration
C_p, C_v	specific heat at constant pressure/volume
D	drag force, strain, duct diameter
De	Deborah number
$\mathcal{D}$	domain or time derivative of a tensor
e	unit vector, wall thickness
E	electric field, modulus of elasticity
$\mathcal{E}$	internal energy
f	function
F	force
g	gravitational acceleration
G	Hooke's constant
h	enthalpy, height of wave
H	hamiltonian
$\mathcal{H}$	Bernoulli sum
J	jacobian
l	characteristic length, length of pendulum
L	characteristic length
$\mathcal{L}$	line in three-dimensional space
m	mass flow
M	Mach number
M_w	molecular weight
Ma	Margoulis number
n	normal direction
N	Brunt–Väisälä frequency
p	pressure
P	running point
$\mathcal{P}$	rate of work
Pr	Prandtl number
q	heat flux per unit area, speed
Q	heat, arbitrary quantity
r	cylindrical radial coordinate, separation distance

R	correlation
$\mathscr{R}$	reference frame
Re	Reynolds number
Ri	Richardson number
s	curvilinear abscissa
S	entropy, surface
t	time
T	temperature, stress
$\mathscr{T}$	work
u	velocity component (fluctuating) in x direction
U	velocity component (mean) in x direction
v	velocity component (fluctuating) in y direction
V	velocity component (mean) in y direction
$\mathscr{V}$	volume
w	velocity component (fluctuating) in w direction
W	velocity component (mean) in w direction
x	cartesian coordinate
y	cartesian coordinate
z	cartesian coordinate
α	coefficient, angle
γ	shear, gas constant
Γ	circulation
δ	unit tensor, boundary layer thickness
∂	partial differential
ε	alternating (Levi–Civita) tensor, energy dissipation, divider opening
ξ	vorticity component in x direction
η	vorticity component in y direction, Kolmogorov scale
ζ	vorticity component in z direction
θ	temperature, tangiantial coordinate
λ	fluid viscosity, Lyapunov exponent
μ	dynamic viscosity, magnetic permeability
ν	kinematic viscosity
π	3.14159 . . .
ρ	density
τ	time, time delay, stress tensor
ϕ	velocity potential
Φ	dissipation function
ψ	stream function
ω	frequency, vorticity
Ω	vorticity

SUBSCRIPTS

∞	far field
e	free stream
o	initial value
c	critical, convection
f	fictitious
g	gas, deficit
H	Hausdorf–Besicovitch
i	running index
l	liquid, laminar
m	maximum,
n	normal
t	turbulent
w	wall
x	at position x

SUPERSCRIPTS

—	time mean
′	differentiation, turbulent fluctuation
·	time derivative
+	law of the wall
~	wave variable
H	heat

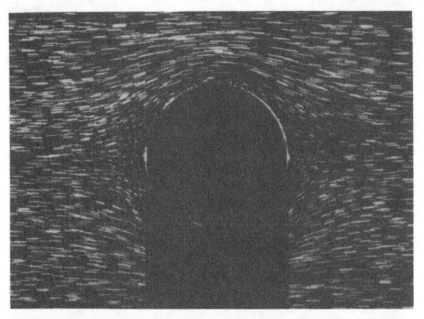

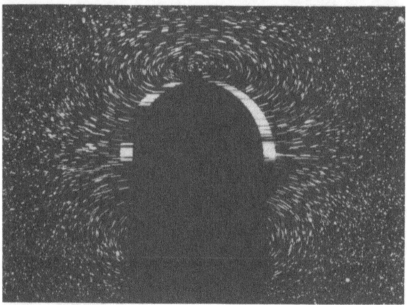

Sphere moving through a tube at Re = 0.10. *Top*, relative motion. *Bottom*, absolute motion. Photograph by M. Coutanceau. Reproduced by permission from Gauthier-Villars, Paris.

1　Basic Concepts

1.1 CONTINUUM

The continuum assumption, used in most physical sciences, presumes that all physical properties of the material under consideration are distributed throughout space. Such an assumption implies that matter is distributed in such a way that the properties of the smallest element into which we can conceive of this matter are the same as those of the substance in bulk.

Unlike classical mechanics, which makes use of point masses, fluid mechanics uses the concept of a continuum almost exclusively. The physical properties of a fluid such as density and flow properties such as velocity can be described using this concept, as we illustrate by means of the classical example of the density of a fluid at a specific point in space at a particular time. The density can be defined as the limit, if it exists, of the ratio of the mass of all molecules (smallest quantities that still retain all the characteristics of the matter under consideration) within an elementary volume about that point in space to the corresponding volume as the volume size decreases to zero.

$$\rho = \lim_{l \to 0} \frac{\Sigma \, m_i}{V} \tag{1.1}$$

Clearly, when the characteristic length l of the representative volume is large, the aforementioned ratio represents an average over a large extent of space, and if the fluid is not homogeneous, this ratio does not remain constant as the length decreases. As the volume decreases, however, there exists a range in which the value of the ratio remains constant. Beyond this range, as the characteristic length approaches intermolecular distances, we can imagine that the value of the ratio will fluctuate (Fig. 1.1). As the characteristic length l of the volume under consideration becomes small enough but before the molecular structure of the matter comes into the picture, the value of the density at the point under consideration is that given by the ratio.

1.2 DESCRIPTION OF FLUID MOTION

Fluid motion can be described in two important ways. The one we are used to from classical mechanics is the lagrangian description, in which

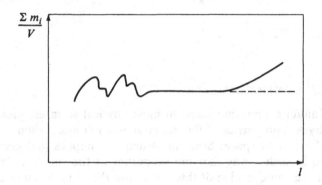

Fig. 1.1   Density variation with characteristic length

we follow one particle. If we choose this method in fluid mechanics, we must follow many particles, each of which must be identified. If $\mathbf{a}$ is the position of a particle at time $t = t_0$, then we can identify this particle as $\mathbf{a}$. It is the name or "color" of the particle. With this definition in mind, the position $\mathbf{x}$ assumed by the particle at a subsequent time t is

$$\mathbf{x} = F(\mathbf{a}, t) \tag{1.2}$$

with

$$\mathbf{a} = F(\mathbf{a}, t_0) \tag{1.3}$$

and the corresponding value of the velocity is

$$\mathbf{V} = \left(\frac{\partial \mathbf{x}}{\partial t}\right)_{\mathbf{a}} \tag{1.4}$$

But if we look at what happens at a fixed location as time passes, we describe the fluid motion in a completely different manner, which is far more common in fluid mechanics and has been named after Euler. In this system, the velocity at a point $\mathbf{x}$ can be expressed as

$$\mathbf{V} = f(\mathbf{x}, t) \tag{1.5}$$

This velocity at point $\mathbf{x}$ and time t must be the same as the velocity of that particle $\mathbf{a}$ at the location $\mathbf{x}$ at time t. We therefore have the following set of three simultaneous differential equations of the first order showing the dependence of $\mathbf{x}$ on t.

$$\frac{\partial \mathbf{x}}{\partial t} = f(\mathbf{x}, t) \tag{1.6}$$

The solution of this set of equations on x involves three constants of integration that satisfy the requirement that at $t = t_0$ the coordinates of the particle become $\mathbf{x} = \mathbf{a}$. From the equations of Lagrange we have

$$\mathbf{x} = F(\mathbf{a}, t) \tag{1.7}$$

so any quantity Q defined in an eulerian system by

$$Q = Q(\mathbf{x}, t) \tag{1.8}$$

can be expressed in a lagrangian system as

$$Q = Q[\mathbf{x} = F(\mathbf{a}, t), t] \tag{1.9}$$

Conversely, if we know the quantity in a lagrangian frame of reference, we can express the corresponding eulerian quantity as

$$Q = Q(\mathbf{a}, t) \tag{1.10}$$

$$\mathbf{x} = F(\mathbf{a}, t) \rightarrow \mathbf{a} = g(\mathbf{x}, t) \tag{1.11}$$

so

$$Q = Q[\mathbf{a} = g(\mathbf{x}, t), t] \tag{1.12}$$

1.3   PATHLINES, STREAKLINES, AND STREAMLINES

If $\mathbf{x} = F(\mathbf{a}, t)$ is the equation in lagrangian form of the particle trajectory in parametric form, then for each t there is a corresponding unique position in space. Such a trajectory is called a *pathline* in fluid mechanics; and as a varies, we get a whole family of such lines, eventually covering the whole flow field.

Just as we can describe a particle trajectory as a function of $\mathbf{a}$ and t we can describe any quantity Q which would be the quantity seen by an observer moving on the trajectory with the particle. The value of Q at the position $\mathbf{x}$ at time t is $Q(\mathbf{x}, t)$, which can be expressed in a lagrangian system as

$$Q = Q[\mathbf{a}(\mathbf{x}, t), t] \tag{1.13}$$

which is the value appropriate to particle a at time t, $\mathbf{a}(\mathbf{x}, t)$ having been obtained by inverting

$$\mathbf{x} = x(\mathbf{a}, t) \tag{1.14}$$

This inversion is always possible if the jacobian of the transformation

$$J = \frac{\partial \mathbf{x}}{\partial \mathbf{a}} \tag{1.15}$$

is not zero. Conversely, we can write

$$Q(\mathbf{a}, t) = Q[\mathbf{x}(\mathbf{a}, t), t] \tag{1.16}$$

A line that is tangent at any instant to the velocity is termed a *stream-*

line, which can be described by writing that the cross product of the velocity by an elementary segment (vector) of the streamline is null:

$$\mathbf{V} \times d\mathbf{r} = 0 \qquad (1.17)$$

which in its extended form is written as

$$\frac{dx}{u} = \frac{dy}{v} = \frac{dz}{w} = ds \qquad (1.18)$$

The equations of *pathlines,* which are lines described by individual particles (i.e., trajectories of particles), are readily available from lagrangian descriptions. One pathline is a line corresponding to the trajectory of a single fluid particle that is identified by some means (usually as described earlier). The equation that describes pathlines is

$$\frac{d\mathbf{x}}{dt} = \mathbf{V}(\mathbf{x}, t) \qquad (1.19)$$

which in its exended form is

$$\frac{dx}{dt} = u \qquad \frac{dy}{dt} = v \qquad \frac{dz}{dt} = w \qquad (1.20)$$

The third important fluid-flow line is the *streakline,* which is the locus at any instant of all particles that have at some previous time passed through a fixed point in the flow field. The equation of a streakline at time t is obtained by considering that a particle is on a streakline if it has passed through a point, say $\mathbf{y}$, at some previous instant between 0 and t. Let's say that the particle was at $\mathbf{y}$ at time τ. The material coordinate of the particle (its name) is then

$$\mathbf{a} = \mathbf{a}(\mathbf{y}, \tau) \qquad (1.21)$$

and at the present time t, this particle is at $\mathbf{x}$ given by

$$\mathbf{x} = \mathbf{x}(\mathbf{a}, t) \qquad (1.22)$$

So the equation of the streakline at time t is

$$\mathbf{x} = \mathbf{x}[\mathbf{a}(\mathbf{y}, \tau), t] \qquad 0 \le \tau \le t \qquad (1.23)$$

For steady flow, needless to say, pathlines, streamlines, and streaklines are identical.

1.4 EXAMPLE OF FLOW LINES

To give a concrete example of flow lines, we shall determine the three lines used for the flow field given by (eulerian form)

$$v_1 = \frac{x_1}{1 + t} \qquad v_2 = x_2 \qquad v_3 = 0 \qquad (1.24)$$

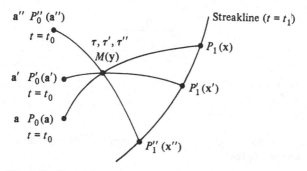

Fig. 1.2   Streakline

The streamlines are solutions of

$$\frac{dx_1}{x_1}(1 + t) = \frac{dx_2}{x_2} = \frac{dx_3}{0} = ds \qquad (1.25)$$

$$x_1 = Ae^{s/(1+t)} \qquad x_2 = Be^{s} \qquad x_3 = \text{constant} \qquad (1.26)$$

These solutions represent lines in the plane $x_3 = c$ having for equation

$$\frac{x_2}{B} = e^{s} = \left(\frac{x_1}{A}\right)^{1+t} \qquad (1.27)$$

The pathlines for this flow are given by

$$\frac{dx_1}{dt} = \frac{x_1}{1 + t} \qquad \frac{dx_2}{dt} = x_2 \qquad \frac{dx_3}{dt} = 0 \qquad (1.28)$$

which again (since there is no motion in the x_3 direction) are confined to lines in $x_1 x_2$ planes. The equations of these pathlines in parametric form (t is the parameter) are given as

$$x_1 = A(1 + t) \qquad x_2 = Be^{t} \qquad x_3 = C \qquad (1.29)$$

where A, B, and C are the particle's a_1, a_2, and a_3 as identified by its position at time $t = 0$. We now look for the equations that describe a streakline consisting of all particles that have passed point y at some previous time τ. The corresponding particles are

$$a_1 = \frac{y_1}{1 + z} \qquad a_2 = \frac{y_2}{e^{\tau}} \qquad a_3 = y_3 \qquad (1.30)$$

Thus, at time t, the equations of the streakline are

$$x_1 = \frac{y_1}{1 + z}(1 + t) \qquad x_2 = y_2 e^{t-\tau} \qquad x_3 = y_3 \qquad (1.31)$$

To get a better idea of what a streakline is, consider three particles a, a', a'' which at times τ, τ', τ'' pass the point M of coordinates y in the

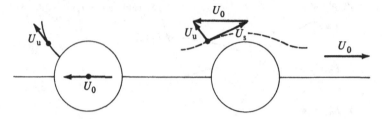

Fig. 1.3   Steady and unsteady flow corresponding to the cases shown
in the photograph

flow field (Fig. 1.2). At a time $t_1 > 0$, the three particles a, a', and a'',
which were at P_0, P_0', and P_0'' at time t_0 are now at P_1, P_1', and P_1'' of
coordinates x, x', and x'' at time $t = t_1$.

1.5 VISUALIZATION OF STREAMLINES, PATHLINES, AND STREAKLINES

Streamlines, pathlines, and streaklines in unsteady flows are generally
distinct, and the interpretation of the flow field by visualization of one
or several of them is a matter of great importance in fluid mechanics.

Whereas pathlines are usually visualized by recording the locations
of particles (properly identified) as time passes, streamlines are best
seen by recording the trace of particles for short periods of time (so
that, at any instant the path described by particles can be assimilated
to the tangent to the streamlines that are collinear with the velocity).

A striking example of visualization of streamlines is provided by Figure 1.3, which depicts patterns of both steady and unsteady flow around
a circular cylinder. Both patterns correspond to the same flow; only the
reference frame is changed. When both the camera and the cylinder are
moved together, the velocity at any point in this frame of reference does
not change, and the flow is steady. On the other hand, when the camera
and the fluid are moved together relative to the cylinder, then the
velocity at any point in the flow field changes with time, corresponding
to unsteady motion. The velocity-vector diagram, with the addition of
a constant velocity that corresponds to relative motion, enables us to
convert from one system to the other. Figure 1.3 illustrates how we
express the correspondence between the unsteady velocity U_u and the
steady velocity U_s. By adding the translational velocity U_0 to the steady
velocity, we obtain the unsteady velocity U_u.

Streaklines are routinely obtained in the laboratory by photographing
dye or smoke released at a point in a flow field. These patterns can be
constructed graphically from the streamline pattern, as shown in Figure
1.4.

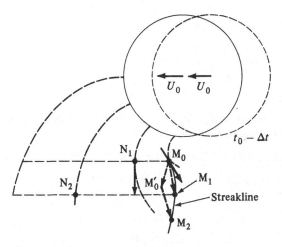

Fig. 1.4   Construction of a pathline

We want to draw a streakline through point M_1 at $t = t_0$. At time $t_0 - \Delta t$, the particle that was at M_0 has moved to M_1. At time $t_0 = 2\,\Delta t$ the particle that was at M_0 has moved to M_2 in such a way that $M_0 M_0'$ is parallel to the velocity at N_1 and $M_0' M_2$ is parallel to the velocity at N_2, and so on.

1.6   BASIC EQUATIONS FOR AN INVISCID FLUID

The state of the fluid is characterized in an eulerian frame by its

velocity	$V(x, y, z, t)$ of components u, v, w
density	$\rho(x, y, z, t)$
temperature	$T(x, y, z, t)$

It is usual in the first approximation to neglect the tangential stresses (Fig. 1.5), so the contact force exerted by the surrounding matter at the surface S bounding the volume is only due to pressure $p(x, y, z, t)$:

$$\iint_s - p\mathbf{n}\, dS$$

The rheological behavior of such a fluid is very simple. But as we shall show, the nature of the contact force is usually more complex and is given in tensorial forms $\mathsf{T}\,(T_{ij})$. Here we have

$$T_{ij} = -p\delta_{ij} \tag{1.32}$$

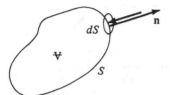

Fig. 1.5 Definition sketch, wherein **n** is the normal to the small element *dS* belonging to *S*

The problem can be solved through the following set of equations.

1. A molecular equilibrium state that is defined by the equation of state. For a perfect fluid, this is expressed as

$$f(p, \rho, T) \tag{1.33}$$

$$p = \rho RT \tag{1.34}$$

in which the pressure term is identified with the pressure term intervening in contact forces.

2. The requirement of the conservation of mass of the fluid, which imposes certain restrictions on the velocity field. Although these restrictions are not strictly kinematic, it is important to introduce them at this stage.

$$\frac{\partial \rho}{\partial r} + \text{div } \rho \mathbf{V} = 0 \tag{1.35}$$

3. When assumptions are made for the rheological behavior of the fluid, the usual momentum equation becomes the Euler equation

$$\mathbf{J} = \frac{d\mathbf{V}}{dt} = \mathbf{F} - \frac{1}{\rho} \text{grad } p \tag{1.36}$$

in which **F** represents the external body forces and **J** the material derivative of the velocity.

$$\frac{du}{dt} = \frac{\partial u}{\partial t} + u\frac{\partial u}{\partial x} + v\frac{\partial u}{\partial y} + w\frac{\partial u}{\partial z} = f_x - \frac{1}{\rho}\frac{\partial p}{\partial x} \tag{1.37}$$

$$\frac{dv}{dt} = \frac{\partial v}{\partial t} + u\frac{\partial v}{\partial x} + v\frac{\partial v}{\partial y} + w\frac{\partial v}{\partial z} = f_y - \frac{1}{\rho}\frac{\partial p}{\partial y} \tag{1.38}$$

$$\frac{dw}{dt} = \frac{\partial w}{\partial t} + u\frac{\partial w}{\partial x} + v\frac{\partial w}{\partial y} + w\frac{\partial w}{\partial z} = f_z - \frac{1}{\rho}\frac{\partial p}{\partial z} \tag{1.39}$$

$$J_i = \frac{du_i}{dt} = \frac{\partial u_i}{\partial t} + u_j\frac{\partial u_i}{\partial x_j} = f_i - \frac{1}{\rho}\frac{\partial p}{\partial x_i} \tag{1.40}$$

The existence of an acceleration potential requires that

$$\mathbf{f} = \text{grad } U \tag{1.41}$$

The density ρ is either constant or a well-defined function of p. (This relationship depends on the properties of the flow under consideration; do not confuse this with the gas law $\mathbf{f}(p, \rho, T)$, which is correlated with molecular properties of the medium.)

$$\text{If} \quad p = p(\rho) \quad \text{and} \quad q(p) = \int_{p_0}^{p} \frac{dp}{\rho} \tag{1.42}$$

then

$$\frac{\partial q}{\partial x} = \frac{dq}{dp}\frac{\partial p}{\partial x} = \frac{1}{\rho}\frac{\partial p}{\partial x} \Rightarrow \text{grad } q = \frac{1}{\rho}\text{grad } p \tag{1.43}$$

The right-hand side of Eq 1.40 can be expressed as

$$\mathbf{J} = \text{grad } U - \text{grad } q \tag{1.44}$$

$$\text{grad}[U - q] = \text{grad } Q \tag{1.45}$$

$$\mathbf{J} = \text{grad } Q \tag{1.46}$$

On the other hand, the left-hand side of Eq 1.40 can be written as

$$\mathbf{J} = \frac{\partial \mathbf{V}}{\partial t} + 2\,\mathbf{\Omega} \times \mathbf{V} + \text{grad}\,\frac{1}{2}\,V^2 \tag{1.47}$$

with

$$\mathbf{\Omega} = \frac{1}{2}\,\text{rot } \mathbf{V} \times \mathbf{V} \tag{1.48}$$

Therefore, comparison of Eq 1.46 with Eq 1.48 leads to

$$\frac{\partial \mathbf{V}}{\partial t} + 2\,\mathbf{\Omega} \times \mathbf{V} = \text{grad}\left[Q - \frac{1}{2}\,V^2\right] \tag{1.49}$$

with

$$H = Q - \frac{1}{2}\,V^2 = U - \int_{p_0}^{p}\frac{dp}{\rho} - \frac{1}{2}\,V^2 \tag{1.50}$$

For steady flow,

$$\frac{\partial \mathbf{V}}{\partial t} = 0 \tag{1.51}$$

Either $\mathbf{\Omega} = 0$ or we go along a streamline because the component of $\mathbf{\Omega} \times \mathbf{V}$ along a streamline is null. In both cases H becomes a constant of the motion.

If we treat the problem as a one dimensional, this constant of motion, supplemented by the continuity equation, enables us to give p and $\mathbf{V}$ at each station. For compressible flows, a connection between p and ρ is required (or a thermodynamic concept such as, for instance, the first law of thermodynamics):

$$H = U - \int_{po}^{p} \frac{dp}{\rho} - \frac{V^2}{2} \qquad (1.52)$$

If gravitational forces are applied, $U = -gz$,

$$-H = gz + \int_{po}^{p} \frac{dp}{\rho} + \frac{V^2}{2} \qquad (1.53)$$

For an incompressible (ρ independent of p) fluid we have

$$-\frac{H}{g} = z + \frac{p}{\rho g} + \frac{V^2}{2g} = \text{constant} \qquad (1.54)$$

If inertial forces are greater than body forces,

$$\frac{p}{\rho g} + \frac{V^2}{2g} = \text{constant} \qquad (1.55)$$

$$p + \frac{1}{2} \rho V^2 = \text{constant} \qquad (1.56)$$

From many assumptions a very simple relation is established between p and $\mathbf{V}$. This relation is purely local, so we have to insist on the restrictiveness of this situation, namely, steady flow of an inviscid, incompressible fluid with negligible body forces.

As a general rule, the pressure is related to the entire velocity field through Poisson's equation. This relationship is discussed in detail later.

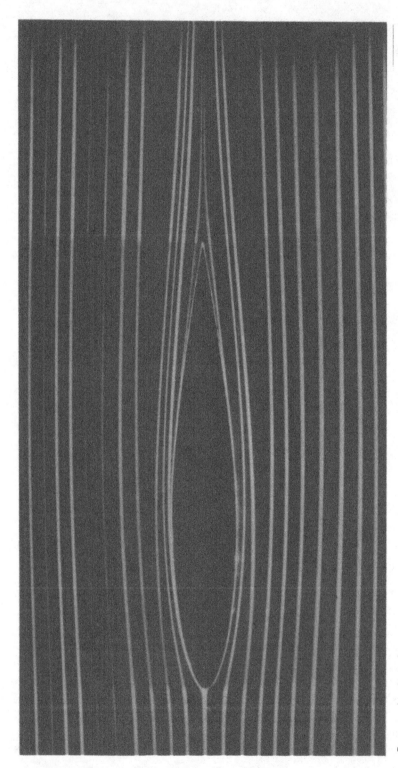

Symmetric plane flow past an airfoil. Photograph by H. Werlé, 1974. Reproduced by permission from Office National d'Etudes et de Recherches Aéronautiques.

2 Toward a Constitutive Law for Fluid Materials

2.1 INTRODUCTION

Before describing the motion of a fluid particle (or a fluid volume) we have to determine what kinds of forces are acting on it. Volume forces and surface forces, which are called long-range forces and short-range forces, respectively, are the two broad classes of forces that act on fluids.

Although we are well acquainted with long-range forces such as body forces induced by the gravity field, surface forces require a brief examination of the rheological behavior of materials. Rheology is "the science of deformation and flow of matter." More precisely, rheology is concerned with those properties of matter that determine how it deforms or flows when subjected to external forces, which may be either inertial or body forces. From this point of view, rheology is considered a material science. On the other hand, if a macroscopic description of the material is sought, rheology can also be considered a branch of mechanics.

Rheological properties are determined by applying the basic laws of mechanics, in particular those that describe the surface forces that represent "force concentration" with dimensions of force per unit area. This rheological approach as presented here is restricted to some fundamental patterns of flows.

2.2 VOLUME FORCES AND SURFACE FORCES ACTING ON A FLUID

Long-range forces such as gravity penetrate the interior of a fluid, acting on all its elements. Gravity is the most obvious and important long-range force, but fictitious forces such as centrifugal forces, which appear to act on mass elements when their motion is referred to a set of accelerating axes, can be treated in the same way as gravity. Electromagnetic forces come into play in some cases: They may act when a fluid carries an electric charge or when an electric current passes through it. Long-range forces decrease slowly with increased distance between interacting elements.

When we write the equations of motion in their general form, the resultant of all body forces acting at time t on the fluid within an element of volume dV surrounding the point whose position is x will be denoted by

$$f(x, t)\rho \, dV$$

The factor ρ, which represents the fluid density, is included because the common type of body forces per unit volume (gravity and the fictitious forces arising from the use of accelerating axes) are proportional to the mass of the element on which they act. In the case of the earth's gravitational field, the force per unit mass is $f = g$; where g acts vertically downward and is time independent.

Short-range forces are molecular and decrease rapidly with increased distance between interacting elements. They are negligible unless there is direct mechanical contact between the interacting elements, as in the case of the reaction between two rigid bodies. We shall see, for instance, that the short-range forces exerted between two masses of gas in direct contact at a common boundary are due predominantly to the transport of momentum across their common boundary by migrating molecules.

Short-range forces can act only on a thin layer of fluid that is adjacent to the boundary of the fluid element acted upon. If the penetration depth of the short-range forces is small compared with the linear dimensions of the plane-surface element, the total force exerted across the element is proportional to its area dS. The total force value at time t acting on an element at position x is represented by the vector

$$T(n, x, t) \, dS$$

At point x, the orientation of the fluid element is given by the unit vector normal to the element n directed toward the outside of the fluid volume under consideration. The force per unit area T is called the *local stress*. The way in which local stress depends on the unit vector that is normal to the surface element is determined later.

2.3   REPRESENTATION OF SURFACE FORCES BY THE STRESS TENSOR

The local state of stress at time t for an element at position x is completely determined if the local stress T is known for any orientation of the unit normal n.

Let us consider all the forces acting on the fluid within an element of volume dV in the shape of a tetrahedron as shown in Fig. 2.1.

We first note that the local body force on the fluid within the tetrahedron is proportional to the volume dV, which is of a smaller order of magnitude than dS. The mass of fluid in the tetrahedron is of the same

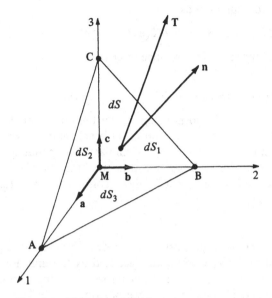

Fig. 2.1   Element of fluid

order of magnitude as V and so is the product of the mass and the acceleration of the fluid in the tetrahedron provided that the local density and acceleration are finite. Thus, if the linear dimension l of the tetrahedron goes to zero without changing its shape, the first two terms of the equation

mass acceleration = resultant of body forces

+ resultant of surface forces

approach zero as the cube of this dimension, whereas the third term approaches zero as the square of it. Accordingly, the equation can be satisfied only if the coefficient of dS vanishes identically. That is:

$$\mathbf{T(n)}\, dS + \mathbf{T(-a)}\, dS_1 + \mathbf{T(-b)}\, dS_2 + \mathbf{T(-c)}\, dS_3 = 0 \qquad (2.1)$$

the three orthogonal faces having for areas dS_1, dS_2, and dS_3, as indicated on the figure and for unit outward normal:

$$-\mathbf{a} \qquad -\mathbf{b} \qquad -\mathbf{c}$$

(The dependence of $\mathbf{T}$ on $\mathbf{x}$ and t is not explicitly indicated here.) The fourth inclined face has an area dS and unit normal $\mathbf{n}$. Projection of dS onto the three orthogonal faces yields

$$dS_1 = \mathbf{a} \cdot \mathbf{n}\, dS = a_j n_j\, dS \qquad (2.2)$$

Equation 2.1 can be expressed as

$$\{\mathbf{T(n)} = (a_j\mathbf{T(a)} + b_j\mathbf{T(b)} + c_j\mathbf{T(c)})n_j\} \tag{2.3}$$

The i component of the sum of the surface forces can therefore be written as

$$\{T_i(\mathbf{n}) = (a_jT_i(\mathbf{a}) + b_jT_i(\mathbf{b}) + c_jT_i(\mathbf{c}))n_j\} \tag{2.4}$$

Herein, the vectors $\mathbf{n}$ and $\mathbf{T}$ do not depend on the choice of the axes of reference, and the expression within curly brackets must represent the (i, j) component of a quantity that is similarly independent of the axes. In other words, the expression within the curly bracket is one component of a second order tensor $\mathbf{T}$ whose components are T_{ij}.

This finally becomes

$$T_i = T_{ij}n_j \tag{2.5}$$

Thus, the components of the stress in the i direction across a plane surface element of arbitrary orientation specified by the unit normal $\mathbf{n}$ is related to the same component of stress across any three orthogonal plane surface elements located at the same position in the fluid and at time t.

Accordingly, the state of stress at a given point is represented by a complex mathematical entity called tensor $\mathbf{T}$. In a three-dimensional space, this second-order tensor has nine components T_{ij}. In the tensorial space the system of reference is

$$\mathbf{e}_i \otimes \mathbf{e}_j \tag{2.6}$$

$$\mathbf{T} = T_{ij}\,\mathbf{e}_i \otimes \mathbf{e}_j \tag{2.7}$$

$$\mathbf{T} = T_{ij}^*e_i^* \otimes e_j \tag{2.8}$$

$T_{11}\,dS_1$, $T_{22}\,dS_2$, $T_{33}\,dS_3$ are the three normal stresses corresponding respectively to the three orthogonal surface elements:

$$OBC \rightarrow T_{11}$$
$$OCA \rightarrow T_{22}$$
$$OAB \rightarrow T_{33}$$

For an inviscid fluid, only normal stresses exist and $T_{11} = T_{22} = T_{33}$, so the second-order tensor takes the simple form $T_{ij} = -p\delta_{ij}$. The tensor is completely defined by the scalar quantity p.

Let us consider now the moment of the various forces acting on the fluid within a volume V of arbitrary shape. The i component of the total moment about a point O may be written as

$$\iint_S \varepsilon_{ijk}OM_jT_k\,dS + \iiint_V \varepsilon_{ijk}OM_jf_k\,dV = \iiint \varepsilon_{ijk}OM_jJ_k\,dV \tag{2.9}$$

The surface forces come into play through the first term. OM_j is a component of the position vector **OM** of the surface element relative to O. The integral over a closed surface can be transformed by the divergence theorem to the volume integral. Since

$$T_k = n_l T_{lk}$$

then

$$\iint_S \varepsilon_{ijk} OM_j T_{lk} n_l \, dS$$

$$= \varepsilon_{ijk} \iiint_V \frac{\partial OM_j}{\partial x_l} T_{lk} \, dV + \varepsilon_{ijk} \iiint_V OM_j \frac{\partial T_{lk}}{\partial x_l} \, dV \qquad (2.10)$$

Recall the divergence theorem

$$\iint A \cdot d\sigma = \iiint \text{div } A \, dV \qquad (2.11)$$

Equation 2.9 then becomes

$$\varepsilon_{ijk} \iiint_V \left\{ T_{jk} + OM_j \left(\frac{\partial T_{lk}}{\partial x_l} + f_k - J_k \right) \right\} dV = 0 \qquad (2.12)$$

This equation can be verified for any arbitrary volume V because

$$\frac{\partial OM_j}{\partial x_l} = \frac{\partial x_j}{\partial x_l} = \delta_{jl} \qquad \text{and} \qquad \delta_{jl} T_{lk} = T_{jk}$$

The fundamental law of mechanics applied to the center of gravity of the fluid element gives

$$\frac{\partial T_{lk}}{\partial x_l} + f_k = J_k \qquad (2.13)$$

Accordingly, the underlined quantity is null. The nine components T_{ij} of the stress tensor **T** are not independent and satisfy the following relationship (note that properties of a polarized dielectric medium could invalidate the second fundamental relation):

$$\varepsilon_{ijk} T_{jk} = \frac{1}{2} \varepsilon_{ijk} (T_{jk} - T_{kj}) = 0 \qquad (2.14)$$

$$T_{jk} = T_{kj} \qquad (2.15)$$

The stress tensor is symmetrical and has only six independent components. This property of the stress tensor is illustrated very simply in two dimensions as shown in the figure.

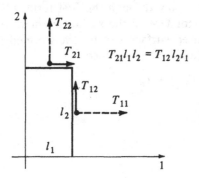

Remarks:

1. The symmetrical form of the tensor T_{ij} can be deduced in another way.

The i component of the total moment about a point O within this volume, exerted by the surface stress at the boundary of the volume, is $\iint_S \varepsilon_{ijk} r_j T_{kl} n_l \, dS$, where r is the position vector of the surface element $(\mathbf{n}, \, dS)$ relative to O. This integral over a closed surface can be transformed by the divergence theorem into

$$\iiint_V \varepsilon_{ijk} \frac{\partial [r_j T_{kl}]}{\partial r_l} \, dV = \iiint_V \varepsilon_{ijk} \left\{ T_{kj} + r_j \frac{\partial T_{kl}}{\partial r_l} \right\} dV \qquad (2.16)$$

If now the volume V is reduced to zero in such a way that the configuration made up of the boundary of the volume and the fixed point O retains its shape, the first term on the right side of Eq. 2.16 decreases as V whereas the second term goes to zero more quickly as $V^{4/3}$. Thus, $\iiint_V \varepsilon_{ijk} T_{kj} \, dV$ is apparently of larger order than all the other terms in the moment equation, and consequently it must be identically zero. This is possible for any position O and shape of V, with T_{ij} continuous in $\mathbf{x}$, only when $\varepsilon_{ijk} T_{kj} = 0$. This relation can also be established considering surface forces only. It is also assumed here that the derivatives of the stress tensor $\partial T_{kl}/\partial r_l$ are of the same order of magnitude as the component of the stress tensor $T_{kl} n_l$.

2. Geometrical approach

For symmetrical tensors, to a quadratic form such as $T_{ij} x_i x_j$ can be associated a quadric $T_{ij} x_i x_j = 1$, which has also an intrinsic form as T.

It is possible to choose the directions of the orthogonal axes of reference so that the nondiagonal elements of the second-order tensor are all zero. Referring to such principal axes at a given point $\mathbf{x}$, the diagonal elements of the stress tensor become principal stresses T'_{11}, T'_{22}, T'_{33}.

The "intrinsic structure" of a tensor is independent of the choice of the axes of reference because the shape of the associated quadric is independent of this choice. This shape is first defined by its volume. Reference can be made to the size of an arbitrary principal axis in order to define the sizes of the other two. Finally, three parameters are necessary to define the shape of the quadric. For tensors, invariants can be defined that characterize their intrinsic structures. For instance, it is a well-known property of second-order tensors that changes of directions of orthogonal axes of reference leave the sum of the diagonal elements (the trace) unchanged.

$$T_{11} + T_{22} + T_{33} = T_{ii}$$

The three invariants of a tensor such as T are

$$I \rightarrow T_{ii}$$

$$II \rightarrow \frac{1}{2}\{I^2 - T_{lm}T_{ml}\}$$

$$III \rightarrow \det T_{lm}$$

For the strain tensor

$$D_{lm} = \frac{1}{2}\left\{\frac{\partial U_1}{\partial x_m} + \frac{\partial U_m}{\partial x_l}\right\} \tag{2.17}$$

and the corresponding three invariants can be written as

$$I_D \rightarrow \frac{\partial U_l}{\partial x_l}$$

$$II_D \rightarrow \frac{1}{2}\left\{I - \frac{\partial U_l}{\partial x_m}\frac{\partial U_m}{\partial x_l}\right\}$$

$$III_D \rightarrow \det T_{lm}$$

These invariants are directly connected with the shape of the associated quadric form.

2.4   THE EQUATIONS OF MOTION

The fundamental law, applied to a volume *V*,

mass × acceleration = Σ body forces + Σ surface forces

with the variables defined in the figure, results in the following equation:

$$\iiint_V \rho(\mathbf{x},\,t)\mathbf{J}(\mathbf{x},\,t)\;dV$$

$$= \iiint_V \rho(\mathbf{x},\,t)\mathbf{f}(\mathbf{x},\,t)\;dV + \iint_S \mathbf{T}(\mathbf{x},\,t)\;dS(\mathbf{n}) \qquad (2.18)$$

The i component can be written as

$$\iiint_V \rho J_i\;dV = \iiint_V \rho f_i\;dV + \iint_S T_i\;dS(\mathbf{n}) \qquad (2.19)$$

The component $T_i\,dS(\mathbf{n})$ is obtained from the stress tensor $T_{ij}(\mathbf{x},\,t)$ as $T_i = T_{ij}n_j$. Using the divergence theorem, we can write the aforementioned equation as

$$\iiint_V \rho J_i\;dV = \iiint_V \rho f_i\;dV$$

$$+ \iiint_V \frac{\partial T_{ij}}{\partial x_j}\;dV \quad \text{or} \quad \iiint_V \left\{\rho J_i - \rho f_i - \frac{\partial T_{ij}}{\partial x_j}\right\}\;dV = 0 \qquad (2.20)$$

Because this relation has to be verified for any arbitrary volume V, we must have

$$\rho J_i = \rho f_i + \frac{\partial T_{ij}}{\partial x_j} \qquad (2.21)$$

in which

$$J_i = J_i[\mathbf{x},\,t] \qquad \rho = \rho[\mathbf{x},\,t] \qquad f_i = f_i[\mathbf{x},\,t] \qquad T_{ij} = T_{ij}[\mathbf{x},\,t]$$

If T_{ij} is assumed to be equal to $p\delta_{ij}$, this dynamic relation becomes the Euler equation:

$$\rho J_i = \rho f_i - \frac{\partial p}{\partial x_i}$$

$$T_{ij} = -p\delta_{ij} \qquad (2.22)$$

At this stage, to solve the basic problem of fluid mechanics, we have the following equations.

  a. The mass conservation equation is one of the fundamental equations of fluid mechanics:

$$\frac{\partial \rho}{\partial t} - \frac{\partial}{\partial x_j} [\rho U_j] = 0 \qquad (2.23)$$

b. For the equation of state, which is a consequence of molecular be-havior, the characteristic times of molecular rearrangements are gen-erally small in comparison with the characteristic time of the whole flow (for instance, a time scale linked to the velocity gradient $\partial U_i / \partial x_j$). Therefore, we generally assume the gas to be in a molecular equilibrium state and the molecular mechanisms that come into play to be expressed by the equation of state

$$\rho(p, \rho, \theta) = 0$$

c. The dynamic equation has been given in the form:

$$\rho J_i = \rho f_i + \frac{\partial T_{ik}}{\partial x_k} \qquad (2.24)$$

d. These equations have to be supplemented by the first principle of thermodynamics.

Regarding the unknown functions of **x** and t, we have an open problem because we have fewer equations than unknowns.

$U(\mathbf{x}, t)$	velocity of the fluid particle,	3 unknowns
$\rho(\mathbf{x}, t)$	density of the fluid	1 unknown
$\theta(\mathbf{x}, t)$	temperature	1 unknown
T	symmetrical stress tensor	<u>6 unknowns</u>
		11 unknowns

For a three-dimensional space we have at our disposal six scalar func-tions. Because the problem under consideration is open, we must in-troduce additional assumptions. One of the simplest is to assume the fluid to be inviscid, for which case $T_{ij} = -p\delta_{ij}$. The surface forces are taken into account through a unique scalar function that is termed the *pressure*. The problem becomes closed because we now have six equa-tions to solve a problem involving six unknowns.

2.5 PRELIMINARY REMARKS ON THE INTERNAL PROPERTIES OF CONTINUOUS MEDIA

Two kinds of laws are encountered in fluid mechanics: *conservation laws,* such as the continuity equation, equation of motion, and energy equation, in which all the quantities are of the same nature; and *con-stitutive laws,* which attempt to establish a correspondence between quantities that are not of the same nature, such as the stress tensor T and the strain tensor D. This correspondence involves coefficients that are closely related to the properties of the medium.

Under very restrictive conditions, it is possible to derive the Navier–Stokes equations (general equations describing fluid motion). At this point, however, we do not intend to develop complex mathematical concepts but rather to bring to light the salient physical principles that are involved in dealing with physics and mechanics.

2.5.a *Point-to-point correspondence*

In many physics and mechanics texts, relations are proposed between physical parameters without making reference to basic assumptions concerning the properties of the medium. For instance, where electrostatic phenomena are concerned, we write

$$\mathbf{D}(\mathbf{x},\, t) = \varepsilon \mathbf{E}(\mathbf{x},\, t) \qquad (2.25)$$

in which $\mathbf{E}$ is the electric field.

The medium is characterized by ε. Two assumptions have been made implicitly: The correspondence between $\mathbf{D}$ and $\mathbf{E}$ is linear, and the medium does not introduce any distortion (i.e., the medium exhibits isotropic properties as far as dielectric phenomena are concerned). If a linear dependence is assumed to exist, then the correspondence between vectors $\mathbf{D}$ and $\mathbf{E}$ should be given in the general form

$$D_i = \varepsilon_{ij} E_j \qquad (2.26)$$

Moreover, if the medium is assumed to be isotropic (as to its dielectric properties), then ε_{ij} has the simple form $\varepsilon \delta_{ij}$, and vectors $\mathbf{D}$ and $\mathbf{E}$ are collinear. The medium does not introduce a distortion between the two fields $\mathbf{D}(\mathbf{x},\, t)$ and $\mathbf{E}(\mathbf{x},\, t)$.

The typical form of ε_{ij} for an isotropic medium is found by considering the scalar bilinear form $\varepsilon_{ij} a_i b_j \rightarrow$ invariant, in which $\mathbf{a}$ and $\mathbf{b}$ are two arbitrary vectors associated with ε_{ij}.

In group theory, the product of two invariant quantities with respect to the group of rotation and symmetry is also an invariant, so

$$\varepsilon_{ij} a_i b_i \rightarrow \text{invariant}$$

$$\text{invariant} \times \text{invariant} \rightarrow \text{invariant}$$

The unique invariant form available from the two vectors $\mathbf{a}$ and $\mathbf{b}$ is the scalar product $\varepsilon \delta_{ij} a_i b_j$, ε being a scalar quantity that does not play any particular role at this stage. In general, ε can be identified as a scalar quantity capable of characterizing the dielectric properties of the isotropic medium.

Comparing the two expressions $\varepsilon \delta_{ij} a_i b_j = \varepsilon_{ij} a_i b_j$, we obtain

$$\varepsilon_{ij} = \varepsilon \delta_{ij} \qquad (2.27)$$

A simple example of the linear correspondence between two vectors is illustrated in the figure for the vectors **D** and **E**. It is possible to extend these considerations to tensorial quantities. In this case, it is also possible to consider associated quadratic forms and eigenvectors.

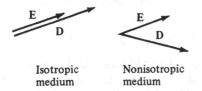

Isotropic   Nonisotropic
medium   medium

Similar properties of the medium are defined when we study thermal conductivity phenomena. The thermal density vector **q** is assumed to be related to the temperature gradient according to Fourier's law.

$$\mathbf{q}(\mathbf{x},\, t) = -\lambda \text{ grad } T(\mathbf{x},\, t)$$

$$q_i = -\lambda \frac{\partial T}{\partial x_i}$$

Such an equation assumes the medium to be isotropic. If we assume a linear relationship, we must use a more general form,

$$q_i = -\lambda_{ij} \frac{\partial T}{\partial x_j} \tag{2.28}$$

from which we can obtain the former expression by writing $\lambda_{ij} = \lambda \delta_{ij}$.

In summary, **D** is a tensorial isotropic function of **E** that can be written as

$$D_i(\mathbf{x},\, t) = f_{ij}[E_j(\mathbf{x},\, t)]$$

Through the tensorial isotropic function, the orientation of **E** is not altered. The correspondence is isotropic through f_{ij} and not through the entities that are corresponding to each other (vectorial quantities **E** and **D**). In any case, a point-to-point correspondence is defined in such a way that the vector **D** at point **x**, t is dependent on the vector **E** at the same point **x**, t.

2.5.b   *Memory effects*

In complex phenomena, a mechanical (or physical) property may depend on a double environment. The stress at a point **x**, t for example, may be connected to the straining effects located in a more or less extended domain (**x** + **r**′); moreover, if memory effects come into play, the straining effects at a previous time $(t - \tau)$ also has to be taken into account. In other words, T (**x**, t) depends on a more or less extended domain of the (x, *t*) space. It is therefore clear that T is a tensorial function of D.

If we conjecture that T is a function of the orientation of D only and not of distorting effects related to the nature of the medium, then the tensorial functional becomes an isotropic tensorial functional, which can be expressed as

$$T_{ij}(\mathbf{x}, t) = \mathcal{F}_{ij}[D_{lm}(\mathbf{x} + \mathbf{r}), t - \theta] \tag{2.29}$$

This mathematical representation does not presume D to be a continuous function of x and *t*. The problem is more tractable if the derivatives exist. We assume that it is so. If, on the other hand, the phenomenon depends on a unique variable, say *t*, memory effects may be introduced through a differential equation. As an example, let us consider Langevin's equation:

$$\frac{dy}{dt} + \beta y = \alpha$$

in which α is a function of time *t*. Originally, Langevin assumed α at β to be random functions of time. The solution then can be written as

$$y(t) = y_0 e^{-\beta(t-t_0)} + \int_{t_0}^{t} e^{-\beta(t-\tau)} \alpha(\tau)\, d\tau \tag{2.30}$$

The first term takes into account the role of initial conditions. We focus on the last term. The exponential term plays the role of a kernel function, which bridges the gap between the two terms in τ and *t*.

At time τ, the disturbance $\alpha(\tau)$ is generated.
At time *t*, its effect $y(t)$ is observed.

Through an exponential form, we take into account vanishing memory effects. The exponential term is, of course, a consequence of the linearity of the basic equation. We can say that *y* (at time *t*) has a functional dependence on α (at time τ).

In a more general form, we can write $y(t)$ as a convolution integral

$$y(t) = \int k(t - \tau)\alpha(\tau)\, d\tau \tag{2.31}$$

Such a convolution product is connected with both the memory effects and the linear dependence.

The spatial dependence can be explained by analogy to the loaded beam problem (see figure). The elementary effects are given by

$$df[x] = K[x, y]p[y]\, dy$$

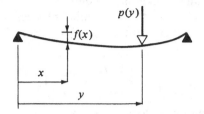

and the global effects obey the integral equation

$$f[x] = \int k[x, y]p(y)\,dy$$

In Langevin's equation, if we assume $\alpha(\tau)$ to be a regular function of τ, then we can expand $\alpha(\tau)$ in a power series so that information at time τ could be deduced from detailed information at time t. In this case, $y(t)$ should become a function of $\alpha(t)$ and of its derivatives $\alpha'(t)$ and $\alpha''(t)$. The functional dependence becomes a simple function of many arguments that are the derivatives of (t) at time t. It is possible to give the solution of a differential equation in a functional form. The reciprocal proposition is not necessarily true; the differential equation obviously requires the existence of derivatives, not the functional form, which supposes the integral to be meaningful.

2.6   THE NAVIER–STOKES EQUATIONS

The Navier–Stokes equations correspond to a simple constitutive law. They represent the first degree of sophistication encountered after the Euler equations. It is remarkable that such simple behavior could be representative of many kinds of fluid flows such as liquids and gases. These equations have numerous applications and can be established in several ways. Here, we start from the momentum equation in its general form.

$$\rho J_i = \rho f_i + \frac{\partial T_{ki}}{\partial x_k} \tag{2.32}$$

To solve the problem under consideration we have to express the tensor T as a function of the kinematic quantities linked to the tensorial gradient of U_l:

$$\frac{\partial U_l}{\partial x_m}$$

We can split this tensorial gradient, like any second-order tensor, into a symmetrical tensor (in connection with a straining process) and an antisymmetrical tensor (in connection with the local rotation of a fluid particle).

$$\frac{\partial U_l}{\partial x_m} = \frac{1}{2}\left[\frac{\partial U_l}{\partial x_m} + \frac{\partial U_m}{\partial x_l}\right] + \frac{1}{2}\left[\frac{\partial U_l}{\partial x_m} - \frac{\partial U_m}{\partial x_l}\right] \tag{2.33}$$

$$\mathcal{D}_{lm} = \underbrace{\text{local straining}}_{D_{lm}} + \underbrace{\text{local rotation}}_{R_{lm}}$$

We seek a relation between T and D and present several arguments to justify this choice. T should be an objective tensor; that is, T should not be affected by a rotation of the axes and thus should depend on objective quantities only. D is an objective quantity not $\mathcal{D}_{lm}$ through R. The stress tensor should be closely connected to local strains rather than to a local body motion. In any case, T is supposed to be dependent on D only.

At this stage, we have not taken memory effects into account. T is immediately and locally adjustable to D. T is a tensorial function of D:

$$T_{ij}(\mathbf{x}, t) = f_{ij}[D_{lm}(\mathbf{x}, t)] \tag{2.34}$$

2.6.a   *A first approach using a quadrilinear form*

If we assume that the relation is local in the $\mathbf{x}, t$ space and the dependence is linear, we can write the stress tensor as

$$T_{ki} = \varepsilon_{kilm}D_{lm} \tag{2.35}$$

In this tensorial operation, a double contraction with respect to l and m is implicated. The rheological properties of the medium are embodied in ε_{kilm}. Moreover, because the medium is assumed to be isotropic, we have to find a typical form for ε_{ij} that should be invariant to both rotation and symmetry.

A quadrilinear form can be associated with ε_{ij} that is invariant to rotation and symmetry; **a, b, c,** and **d** are four arbitrary vectors associated with ε_{ij}.

$$\varepsilon_{kilm}a_ib_kc_ld_m \rightarrow \text{invariant}$$

Starting from four vectors, we introduce the unique invariants to the group of rotation and symmetry, which are the scalar products

$$(\mathbf{a} \cdot \mathbf{b})(\mathbf{c} \cdot \mathbf{d}) \qquad (\mathbf{a} \cdot \mathbf{c})(\mathbf{b} \cdot \mathbf{d}) \qquad (\mathbf{a} \cdot \mathbf{d})(\mathbf{b} \cdot \mathbf{c})$$

By considering only these three quantities, which are useful for identification, we can write

$$\varepsilon_{kilm}a_kb_ic_ld_m = Aa_kb_ic_ld_m\delta_{ki}\delta_{lm} + Ba_kb_ic_ld_m\delta_{kl}\delta_{im} + Ca_kb_ic_ld_m\delta_{km}\delta_{il}$$

After identification we obtain

$$\varepsilon_{kilm} = A\delta_{ki}\delta_{lm} + B\delta_{kl}\delta_{im} + C\delta_{km}\delta_{il} \qquad (2.36)$$

in which A, B, and C are adjustable constants that characterize the intrinsic properties of the medium. The correspondence between T_{ik} and D_{lm} may now be established as

$$T_{ki} = A\delta_{ki}\delta_{lm}D_{lm} + B\delta_{kl}\delta_{im}D_{lm} + C\delta_{km}\delta_{il}D_{lm}$$

with D_{lm} representing the rate-of-strain tensor

$$D_{lm} = \frac{1}{2}\left(\frac{\partial U_l}{\partial x_m} + \frac{\partial U_m}{\partial x_l}\right) \qquad (2.37)$$

The stress tensor is linked to the rate-of-strain tensor through the expression

$$T_{ki} = A\delta_{ki}\frac{\partial U_l}{\partial x_l} + (B + C)D_{ki}$$

This component, which is only a consequence of straining effects, must be supplemented by the pressure term pn, which does not make reference to a straining process.

$$T_{ki} = A\delta_{ki}\frac{\partial U_l}{\partial x_l} + (B + C)D_{ki} - p\delta_{ki} \qquad (2.38)$$

T_{ki} now also denotes the value of the global stress tensor component. We set

$$A = \lambda \qquad B + C = 2\mu$$

Taking the divergence of T_{ki}, we obtain

$$\frac{\partial}{\partial x_k}T_{ki} = -\frac{\partial p}{\partial x_i} + \lambda\frac{\partial}{\partial x_i}\operatorname{div}\mathbf{U} + 2\mu\frac{\partial}{\partial x_k}D_{ki}$$

$$2\mu\frac{\partial}{\partial x_k}D_{ki} = \mu\frac{\partial}{\partial x_k}\left\{\frac{\partial U_k}{\partial x_i} + \frac{\partial U_i}{\partial x_k}\right\}$$

$$\frac{\partial}{\partial x_k}T_{ki} = -\frac{\partial p}{\partial x_i} + (\lambda - \mu)\frac{\partial}{\partial x_i}\operatorname{div}\mathbf{U} + \mu\frac{\partial U_i}{\partial x_k\,\partial x_k}$$

This finally leads to the Navier–Stokes equations

$$\rho J_i = \rho f_i - \frac{\partial p}{\partial x_i} + (\lambda - \mu)\frac{\partial}{\partial x_i}\operatorname{div}\mathbf{U} + \mu\frac{\partial^2 U_i}{\partial x_k\,\partial x_k} \qquad (2.39)$$

which in a vectorial form can be written as

$$\rho\mathbf{J} = \rho\mathbf{f} - \operatorname{grad}p + (\lambda - \mu)\operatorname{grad}(\operatorname{div}\mathbf{U}) + \mu\,\Delta\mathbf{U}$$

2.6.b *An approach using representation theorems*

This method has successfully been used not only in mechanics of continuum media but also to determine the form of typical functions in turbulence fields such as dissipation, rate of strain pressure correlation, and so on. A local relation between $T(X, t)$ and D is still postulated together with thermodynamics concepts. Under these conditions, the stress tensor can be expressed as

$$T_{ij} = f_{ij}(D_{ij})$$

and $T_{ij} = -\pi\delta_{ij}$ when $D_{lm} = 0$ (fluid at rest), where π is a scalar quantity called *thermodynamic pressure*. In fact, we have no way of deducing the dependence of D on T for fluids in general, except by introducing several characteristic parameters.

When the molecular structure of the fluid is statistically isotropic, the orientations of T are the same as those of D, and we can write (by using a representation theorem)

$$f_{ij}(D_{lm}) = \alpha_0\delta_{ij} + \alpha_1 D_{ij} + \alpha_2 D_{ik}D_{kj} \qquad (2.40)$$

where α_0, α_1, and α_2 are scalar functions of the three invariants of the strain tensor. We recall that

$$I_D = \frac{\partial U_l}{\partial x_m}$$

$$II_D = \frac{1}{2}\left\{I_D^2 - \frac{\partial U_l}{\partial x_m}\frac{\partial U_m}{\partial x_l}\right\}$$

$$III_D = \det D_{lm}$$

Stokes's assumption leads to

$$T_{ij} = (-\pi + \alpha_0)\delta_{ij} + \alpha_1 D_{ij} + \alpha_2 D_{ik}D_{kj} \qquad (2.41)$$

with $\alpha_k = \alpha_k(I_D, II_D, III_D)$ and $\alpha_0 = \alpha_0[0, 0, 0] = 0$.

The quantities α_k have to be determined empirically. Moreover, in thermodynamic theories the second principle restricts the choice of these terms because the dissipation has to be negative or null.

To get the Navier–Stokes equations, we make the following assumptions:

$$\alpha_0 = \lambda I_D \qquad \alpha_1 = 2\mu \qquad \alpha_2 = 0$$

where λ and μ are functions of the thermodynamic variables ρ and T only. The stress tensor can then be expressed as

$$T_{ij} = (-\pi + \lambda I_D)\delta_{ij} + 2\mu D_{ij}$$

A necessary condition for the dissipation term to be positive is $3\lambda + 2\mu \geq 0$; in fact, it is usually assumed that $3\lambda + 2\mu = 0$. This assumption is supported by simple inferences. The mechanical pressure is defined by $p = -\frac{1}{3}T_{kk}$ (mean value of the trace of T_{ij}). From the previous law, it is found that

$$-3p = T_{kk} = 3[-\pi + \lambda I_D] + 2\mu I_D$$

$$3[p - \pi] + (3\lambda + 2\mu)I_D = 0$$

For incompressible fluids, $I_D = 0$ and $p = \pi$. For compressible fluids, $p = \pi$ if

$$3\lambda + 2\mu = 0 \tag{2.42}$$

and otherwise, $p \neq \pi$.

2.6.c *Some preliminary remarks on the structure of the Navier–Stokes equations*

The relationship between T and D previously established

$$T_{ik} = A\delta_{ki}\frac{\partial U_l}{\partial x_l} + (B + C)D_{ki} \tag{2.43}$$

appears to be a component-to-component correspondence since it can be inferred from elementary considerations on vectorial correspondence.

Whenever a nondistorting medium is being considered, the axes of symmetry of the two quadrics associated with T_{kv} and D_{lm}, respectively, coincide

$$\{T_{ki}x_kx_i = 1; \qquad D_{lm}x_lx_m = 1\}$$

and the corresponding eigenvectors are then collinear.

In fact, it is sufficient to split the straining tensor D_{ki} between a straining process with constant volume, $D_{ik} - \frac{1}{3}$ div $U\delta_{ik}$ (deviatoric tensor), and a second process that takes into account the evolution of only the elementary volume, δ_{ik} div U. A typical coefficient is associated with each process.

With the Navier–Stokes equations rewritten in the form

$$\rho\left(\frac{\partial u_i}{\partial t} + u_j\frac{\partial u_i}{\partial x_j}\right)$$

$$= \rho f_i - \frac{\partial}{\partial x_i}p + (\lambda + \mu)\frac{\partial}{\partial x_i}\left(\frac{\partial U_l}{\partial x_l}\right) + \mu\frac{\partial^2 U_i}{\partial x_l \partial x_l} \tag{2.44}$$

we can make two observations. First, the nonlinear character of this equation is a consequence of the advective terms–the viscous effects introduce linear effects only–which is a straightforward consequence of the relation introduced between T and D. In a simple pattern of flow with straight streamline (uniform flow), the Navier–Stokes equations correspond to a viscous stress defined by

$$\tau = \mu \frac{dU}{dy}$$

so the balance of viscous effects on a streamline is equal to

$$\frac{d\tau}{dy} = \mu \frac{d^2U}{dy^2}$$

The viscous stress is given by a first gradient approximation.

Second, the new differential equations are second order, whereas Euler's equations are first order. This is a drastic change. At this stage we point out only the role played by the second-order terms when the boundary conditions are considered. The fluid particles adjacent to the wall have the same velocity as the wall (the no-slip condition): $U - U_w = 0$.

Now that we have an expression for the stress tensor, we can explicitly determine the boundary conditions to be used in subsequent mathematical determinations of the velocity distribution in a fluid. We can expect internal discontinuities to be eliminated by viscous effects. Although the tangential component of the velocity is continuous across a material boundary separating a fluid from another medium, we point out that the condition of continuity of velocity is not an exact law but a statement of what can be expected in normal circumstances. The effectiveness of the viscous stress in smoothing a discontinuity in fluid velocity depends on the magnitude of the viscosity, among other factors.

The Navier–Stokes equations, which are a special form of the momentum equations, must be supplemented by the continuity equation, the equation of state, and the energy equation to solve the problem.

We begin by applying the momentum equation to the center of gravity of a fluid particle:

$$\rho \frac{dU_i}{dt} = \rho f_i + \frac{\partial T_{ik}}{\partial x_k}$$

After multiplying by U_i and integrating over a finite volume V we obtain

$$\iiint_V \rho \frac{dU_i}{dt} U_i \, dV = \iiint_V \rho f_i U_i \, dV + \iiint \frac{\partial T_{ik}}{\partial x_k} U_i \, dV$$

The rate at which work is being done on the fluid in the material volume

V is the sum of a contribution

$$\iiint\limits_{V} \rho f_i U_i \, dV$$

from the resultant of body forces and a contribution

$$\iiint\limits_{V} \frac{\partial T_{ki}}{\partial x_k} U_i \, dV$$

from the surface forces. To obtain the action of surface forces at the boundary S of V we write

$$\iiint\limits_{V} \rho \frac{\partial T_{ki}}{\partial x_k} U_i \, dV$$

$$= \iiint\limits_{V} \frac{\partial}{\partial x_k} (T_{ki} U_i) \, dV - \iiint\limits_{V} T_{ki} \frac{\partial U_i}{\partial x_k} \, dV$$

$$= \underbrace{\iint\limits_{S} T_{ki} U_i n_k \, dS}_{\substack{\text{stress work at the} \\ \text{boundary surface}}} - \underbrace{\iiint\limits_{V} T_{ki} \frac{\partial U_i}{\partial x_k} \, dV}_{\text{dissipation}} \qquad (2.45)$$

As in previous sections, $T_{ki} n_k$ represents the i component of the surface force per unit area acting on dS; accordingly, $T_{ki} n_k U_i \, dS$ represents the elementary work per unit time due to this force. The dissipation of mechanical energy represented by the last term is equivalent to an irreversible addition of heat in its effect on the fluid.

If T_{ki} is associated with a velocity gradient, the sign of the last integral is completely determined. This relation between stress and the velocity field does not give rise to difficulties as far as the second principle of thermodynamics is concerned. On the other hand, the constitutive law has to be consistent with the second principle of thermodynamics.

2.7   GENERALIZED NEWTONIAN FLUIDS

The most important property of macromolecular fluid for the industrial chemical engineer is the non-Newtonian viscosity, which changes with the shear rate because the viscosity of some typical fluids can change by a factor of 10, 100, or even 1000. Such enormous changes cannot be ignored in pipe-flow calculations, lubrication problems, design of viscometers, extruder operations, polymer processing calculations, and so on.

The simplest way to approach this phenomenon is to allow the viscosity to vary with shear rate (or with shear stress). This can be done only in an empirical manner. We shall try to reconcile this technical point of view with the previous general approach.

For elementary flow (uniform flow), characterized by

$$U = U(y) \qquad V = 0 \qquad w = 0$$

the early rheologist replaced the first gradient law

$$\tau = \mu \frac{dU}{dy} \tag{2.46}$$

with an empirical law

$$\tau = \eta \frac{dU}{dy}$$

η being a function of either dU/dy or τ (i.e., defined as a flow property rather than a fluid property). Note also that the change in viscosity depends on the magnitude, not the sign, of the shear rate (or shear stress).

2.7.a   *General formulation of the problem*

Assume that the velocity field (see figure) is given in the form $U = U(y)$. The rate-of-strain tensor reduces to

$$D_{ij} = \begin{vmatrix} 0 & \tfrac{1}{2}\dot{\gamma} & 0 \\ \tfrac{1}{2}\dot{\gamma} & 0 & 0 \\ 0 & 0 & 0 \end{vmatrix}$$

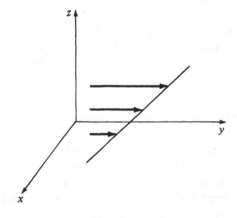

Thus, $\dot{\gamma} = dU/dy$ is the only nonzero component of this tensor. The fluid rheological behavior is decided by the Stokes constitutive equation obtained by representation theorems:

$$T_{ij} = -p\delta_{ij} + \alpha_1 D_{ij} + \alpha_2 D_{ik} D_{kj} \qquad (2.47)$$

In a general approach, α_1, and α_2 are scalar functions of the three principal invariants of the tensor D_{ij}. They can take on the equivalent form:

$$I_D = \text{tr}[D_{ij}]$$

$$II_D = \frac{1}{2}\{\text{tr}[D_{ik}D_{kj}] - (\text{tr}(D_{ij}))^2\}$$

$$III_D = \det(D_{ij})$$

For the case under consideration, these three invariants reduce to

$$I_D = 0 \qquad \text{(incompressible fluid)}$$

$$III_D = 0$$

$$II_D \sim \dot{\gamma}^2$$

The corresponding stress components are

$$T_{11} = T_{xx} = -p + \alpha_2 \dot{\gamma}^2$$

$$T_{12} = T_{xy} = \alpha_1 \dot{\gamma}$$

$$T_{22} = T_{yy} = -p + \alpha_2 \dot{\gamma}^2$$

As direct consequence of preceding first and third equations, it is clear that there is no normal stress difference, $T_{xx} - T_{yy} = 0$, which means in turn that α_2 may be set equal to zero. In this case the three components of the stress tensor become

$$T_{xx} = -p \qquad T_{yy} = -p \qquad T_{xy} = \alpha_1[\dot{\gamma}^2]\dot{\gamma}$$

2.7.b *Illustrative example*

To illustrate the dependence that can exist between the viscosity and the shear rate, we use the simple case of the power-law model

$$T_{xy} = K\{(\dot{\gamma})^2\}\dot{\gamma}^n \qquad (2.48)$$

paying close attention to two cases: $n = 1$ (a pure newtonian fluid for which the classical parabolic profile is obtained) and $n = 0$ (a uniform shear stress and, eventually, a plug-flow). An interesting connection between the two cases can be ascertained by using "yield stress mate-

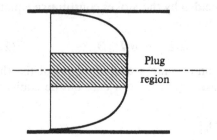

Fig. 2.2   Plug region in pipe flow

rials" such as highly concentrated suspensions and aggregates. The well-known Bingham fluid belongs to this class of fluids.

$$T_{xy} = \eta\dot{\gamma} \quad \text{if} \quad T(x, y) > \tau_0$$

$$\text{either} \quad \dot{\gamma} = 0 \quad \text{or} \quad T(x, y) < \tau_0$$

Plug flow is observed in the central zone in the case of steady shear flow (see Fig. 2.2). The diameter of this plug-flow depends on the magnitude of the yield τ_0.

In the case of a power law $T_{xy} = k\dot{\gamma}^n$ the shape of the velocity profile is given as a function of n (see Fig. 2.3).

2.8   MEMORY EFFECTS

Even though the concept of a generalized Newtonian fluid has proved to be of great value in solving engineering problems, its use is limited to a large extent to steady-state shearing flows. It is generally inappropriate for the description of unsteady flow phenomena in which the elastic response of a polymeric fluid becomes important. In this section, we introduce a precise rheological description of the fluid medium that is capable of accounting for the time-dependent motions of macromolecular fluids. The general linear theory of viscoelastic fluid still assumes flows in which the fluid never moves very far or very rapidly from its initial configuration.

It is by no means necessary to restrict this discussion to a linear phenomenon because we are interested only in an overall view of this subject. The domain that we shall explore can be limited by two extreme cases. For Newtonian fluids, we assume the stress tensor $T(x, t)$ to be immediately adjustable to the local rate-of-strain tensor $D(x, t)$; no memory effects are introduced. This assumption is supplemented by two others: The relationship between the two tensors is linear, and the

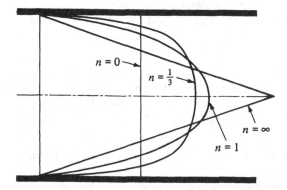

Fig. 2.3 Velocity profiles presented as a function of n (power law)

medium is isotropic. The second case is that of a spring for which the force is continuously dependent on its stretching. More precisely, the force is linked to the displacement in a cumulative way. From this point of view, the behavior of a spring exhibits strong memory effects. Likewise, if we stretch a rubber band and then release it, it springs back to its original length. We say that the rubber is *elastic,* just like the spring. The rubber is a network polymer with permanent chemical cross-links; it can remember its initial unstretched configuration exactly. Polymeric fluids combine the two previous effects. Elastic effects in fluids are more dramatic: In some fluids, when the pressure producing tube-flows is removed, the fluid retreats in the direction from which it came, though it does not return to its original position as the spring does. It seems gradually to forget where it came from.

These remarks are meant merely to introduce notions about elasticity of liquids and fading memories. Though we do not discuss the molecular mechanisms responsible for this macroscopic behavior, we note that the lifetimes of physical junctions may play a role in memory effects. For instance, if the junctions have a lifetime equal to t_0, we can expect the memory of the fluid to extend far beyond t_0. To gain macroscopic insight, we have to introduce ideas about viscosity and elasticity, and in doing so we must mention Maxwell, who speculated over a century ago that gases might be viscoelastic.

Let us look at a one-dimensional model of this second case. We choose a material that exhibits viscoelastic properties and is sandwiched between two planes that are free to move in the x direction. Newtonian behavior is taken into account by introducing a viscosity coefficient μ, whereas hookean solid behavior is characterized by G. The conventional mechanical elements representing linear and elastic behaviors are the spring and the dashpot. We next introduce an idealized model for the rheological behaviors of viscoelastic materials, starting from the two elements represented in the figure. (Time derivatives are denoted by $\dot{x}$ and $\dot{y}$.)

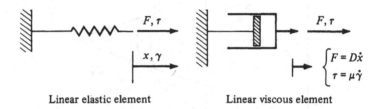

Linear elastic element   Linear viscous element

One of the simplest models that can be derived, known as the Maxwell model, consists of one spring and one dashpot in series (see figure).

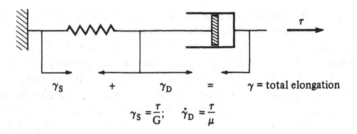

The total strain is the sum of both the elastic strain γ_S of the spring and the fluid strain γ_D of the dashpot, so we can write

$$\gamma = \gamma_S + \gamma_D$$

$$\dot{\gamma} = \dot{\gamma}_S + \dot{\gamma}_D = \frac{\dot{\tau}}{G} + \frac{\tau}{\mu}$$

This one-dimensional model leads to the simple equation

$$\lambda \frac{d\tau}{dt} + \tau = \mu\dot{\gamma} \qquad (2.49)$$

with $\lambda = \mu/G$ being a time constant.

It is instructive to compare this equation with Langevin's equation when memory effects are present. To bridge the gap with complex rheological behavior this equation should be written

$$\frac{d\mathbf{T}}{dt} + \frac{1}{\lambda}\mathbf{T} = \frac{\mu}{\lambda}\mathbf{D} \qquad (2.50)$$

in which $\mathbf{D}$ is the rate-of-strain tensor. This formulation is not too different from the previous one if we suppose a component-to-component correspondence between the two tensors T and D, such as

$$\lambda \frac{dT_{ij}}{dt} + T_{ij} = \mu D_{ij} \tag{2.51}$$

At this stage we do not discuss the meaning of the temporal derivative so that we can present the solution in an integral form:

$$T_{ij}(t) = T_{ij}[t_0]e^{-(t-t_0)/\lambda} + \int_{t_0}^{t} \left\{ \frac{\mu e^{-(t-\theta)/\lambda}}{\lambda} \right\} D_{ij}(\theta) \, d\theta \tag{2.52}$$

All values of the rate-of-strain tensor D in the time interval (t_0, t) interact with the value of the stress tensor T at time θ through a weighing function called the *relaxation modulus*. This function plays the role of a fading memory. Extensions of the Maxwell model can then be made. The physical model is simply a parallel arrangement of many individual Maxwell elements, each exhibiting typical parameters. If we impose the additional condition that $\tau(t_0) = 0$, the solution of the fundamental equation

$$\lambda \frac{d\tau}{dt} + \tau = \mu\gamma$$

with $\lambda = \mu/G$ is presented in the form

$$\tau(t) = \int_0^t G e^{-(t-\theta)/\lambda} \gamma(\theta) \, d\theta$$

or

$$\tau(t) = \int_0^t \Psi(t - \theta)\gamma(\theta) \, d\theta \tag{2.53}$$

The generalized Maxwell model leads us to introduce

$$\Psi(t) = \sum G_i e^{-t/\lambda_i}$$

so we have

$$\tau(t) = \sum_{i=1}^{\infty} G_i \int_0^t e^{-(t-\theta)/\lambda_i} \gamma(\theta) \, d\theta$$

If the spectrum of relaxation times is continuous instead of discrete, then we can write

$$\Psi(t) = \int_0^\infty G(\lambda)e^{-t/\lambda} \, d\lambda$$

therefore,

$$\tau(t) = \int_0^t \int_0^\infty G(\lambda)e^{-(t-\theta)/\lambda}\gamma(\theta)\ d\lambda\ d\theta \qquad (2.54)$$

We interpret $G(\lambda)\ d\lambda$ as the *relaxation modulus*, with a value between λ and $\lambda + d\lambda$, and $G(\lambda)$ as the spectrum of the relaxation modulus. For steady flow, $\dot\gamma$ is independent of time, so we obtain

$$\tau = \dot\gamma \int_0^\infty G(\lambda) \int_{t_0}^t e^{-(t-\theta)/\lambda}\ d\lambda\ d\theta$$

$$= \dot\gamma \int_0^\infty G(\lambda) \int_{t_0}^{t-t_0} e^{-s/\lambda}\ d\lambda\ d\theta$$

$$= \dot\gamma \int_0^\infty \lambda G(\lambda)\ d\lambda$$

That leads to a value of $\tau/\dot\gamma$ given by

$$\frac{\tau}{\dot\gamma} = \int_0^\infty \lambda G(\lambda)\ d\lambda$$

We must integrate along the particle paths. By means of these line integrals, information closely related to the history of fluid elements is introduced. However, the use of material derivatives is suspect if we want to be coherent with the properties required for T. The stress tensor T is expected to be an objective tensor, as is the strain tensor D. If we introduce the material derivatives of D however, this property will be lost. The dependence between T and D can be formulated in a general form by introducing a functional dependence of the form

$$T_{ij} = \mathcal{F}_{ij}(D_{lm}) \qquad (2.55)$$

In so doing we do not presume any relationship between the stress tensor T and the strain tensor D to exist. In any case a local dependence is not expected. However, functional dependences are difficult to handle. This connection could be replaced by a simple tensorial function f_{ij} if new adequate arguments are introduced such as

$$T_{ij} = f_{ij}(D_{lm}, \ldots)$$

These arguments should be closely related to the history of D_{lm}. If derivatives of D_{lm} exist, these arguments should necessarily introduce time derivatives of D_{lm}. We now ask ourselves the question: What kind of derivatives should be chosen? At first, material derivatives would appear to be convenient, but in fact they are not. As previously mentioned, material derivatives are not objective because the operator

$$\frac{D}{Dt} = \frac{\partial}{\partial t} + U_j \frac{\partial}{\partial x_j}$$

introduces vorticity. The acceleration of the fluid particle refers to material derivatives of **U**. The acceleration becomes

$$\mathbf{J} = \frac{\partial \mathbf{U}}{\partial t} + 2\mathbf{\Omega} \times \mathbf{U} + \operatorname{grad} \frac{1}{2} U^2 \qquad (2.56)$$

Vorticity clearly appears in this expression. Disregarding the problem of objectivity, the degree of memory of the fluid depends on the number of arguments taken into account in the expression for the stress tensor:

$$T_{ij} = f_{ij}\left(D_{lm}; \frac{D}{Dt} D_{lm}; \frac{D^2}{Dt^2} D_{lm}; \ldots \right)$$

Fading memory effects appear, which lead to the concept of intricate boundary layers, each exhibiting a typical degree of memory. Some refinements must be introduced to define objective derivatives. Computations are detailed in Appendix A. At this stage, only physical arguments are put forward.

Because the generalized Newtonian model and the linear viscoelastic model are limited in extent, the need to solve dynamics problems involving flows outside these two classes of motion requires us to have a rheological equation of state of greater generality.

According to our previous remarks, the rheological behavior of a fluid material should eliminate the role of a solid rotation. The simplest way is to introduce rheological equations of state into a corotating and cotranslating frame, which should make possible the description of large deformation flows of viscoelastic fluids. Up to this point we have used a fixed frame, the frame in which experimental and computational approaches are usually carried out in the laboratory. The necessity of introducing objective arguments in f_{ij} gives rise to mathematical difficulties, which are outlined in Appendix B.

The corotating frame moves along with the local angular velocity of the fluid defined by $\mathbf{\Omega} = \frac{1}{2} \operatorname{curl} \mathbf{U}$. As shown in the figure, this corotating frame can be chosen so that at time t it is lined up with the fixed frame; for an earlier time $t - \theta = t'$, the corotating frame is tilted. Therefore, the orthogonal unit vectors of this corotating frame are functions of time. An observer in this frame would report stresses in terms of $\mathbf{T}'$ referred to the cotranslating and corotating frame. The corotating observer would then formulate a rheological equation of state between $\mathbf{T}'$ and $\mathbf{D}'$. If the equations of motion are defined in the fixed frame, we

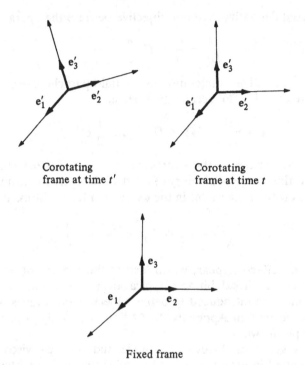

Corotating
frame at time t'

Corotating
frame at time t

Fixed frame

need transformation laws to enable us to change coordinates. The rheological equation of state is transferred from the corotating frame to the fixed frame to validate the equations of motion. Such a transformation necessitates the introduction of the corotational derivatives, also called Jaumann derivatives. The Jaumann derivative of the components of a second tensor T is given in its usual form as

$$\frac{\mathcal{D}}{\mathcal{D}t} T_{ij} = \frac{D}{Dt} T_{ij} + \frac{1}{2} \{\Omega_{ik}(t)T_{kj}(t) - T_{ik}(t)\Omega_{kj}(t)\} \qquad (2.57)$$

This derivative consists of two parts. The usual material derivative D/Dt (also called the substantial or Stokes derivative) and an extra term that accounts for the rotatory motion of the fluid. Jaumann derivatives do not seem to introduce the questionable effects of the surrounding fluid, although this is still an open question. Recent refinements are beyond the scope of this introduction to rheological behavior.

2.9   INFLUENCE OF SCALING ON THE DESCRIPTION OF A PHYSICAL PHENOMENON

Due to differences at the molecular level, the flow behavior of polymeric (macromolecular) fluids is quite unlike that of small-molecule fluids.

Nevertheless, global patterns of the flow can be satisfactorily described by the Navier–Stokes equations. From this point of view, we briefly analyze generalized newtonian fluids, take memory effects into account, and reveal linear and nonlinear effects. The mechanics of continuum media is concerned only with macroscopic approaches; however, the predictions of kinetic theory for the stress tensor of polymeric fluids can also be developed to serve as a useful background. In any description of a mechanical phenomenon a scaling is implicitly adopted. If a statistical description is accepted, new aspects of the problem under consideration are introduced.

A macromolecule is a large molecule composed of many simple chemical units, usually called *structural units*. What happens to these units and to their interrelations when a fluid is subjected to strain? Are some connections broken or not? These internal mechanisms appear globally through some memory effects, which are the consequences of quasi-instantaneous behavior at the molecular scale. We want to emphasize here the importance of the choice of the scaling. For completeness, we shall refer to a subject that is treated in a later chapter. Whenever we encounter a turbulent motion, we use the Navier–Stokes equations to describe the instantaneous motion of the fluid:

$$\frac{\partial U_i}{\partial t} + U_j \frac{\partial U_i}{\partial x_j} = -\frac{1}{\rho}\frac{\partial P}{\partial x_i} + \nu \frac{\partial^2 U_i}{\partial x_j \partial x_j} \tag{2.58}$$

At this stage, ν can be considered a statistical parameter extracted from the kinetic theory of gases. Working with the Navier–Stokes equations, we obtain a detailed description of the fluid motion, which is made at the scale of the fluid particle. No memory effects are introduced through the Navier–Stokes equations. However, if we use a more global description of the turbulent motion, we can introduce a statistical approach. First, we decompose the instantaneous motion into two parts: the mean motion and the fluctuation motion. Using overbars to denote mean values and lower-case letters to denote the fluctuating motion, we have

$$U_i = \overline{U}_i + u_i \qquad P = \overline{P} + p$$

Each statistical moment obeys a partial differential equation that reveals memory effects. The rate equation for the turbulent kinetic energy can be written as

$$\frac{\partial}{\partial t}\overline{q^2} + \overline{U}_k \frac{\partial \overline{q^2}}{\partial x_k} = -\overline{u_i u_k}\frac{\partial U_i}{\partial x_k} - \frac{\partial}{\partial x_k}\overline{q^2 u_k}$$

$$= -\frac{1}{\rho}\frac{\partial}{\partial x_i}\overline{p u_i} + \nu \frac{\partial^2 u_i}{\partial x_k \, \partial x_k} \tag{2.59}$$

This equation describes the motion in a way that is more or less related to the integral scale of turbulence. This scale is more precisely connected with the order of the statistical moment under consideration. In fact, we expect the statistical moments of the highest order to be related to the smallest structures in the fluid. With a statistical approach, turbulence behaves like a kind of material, displaying memory effects and elastic properties.

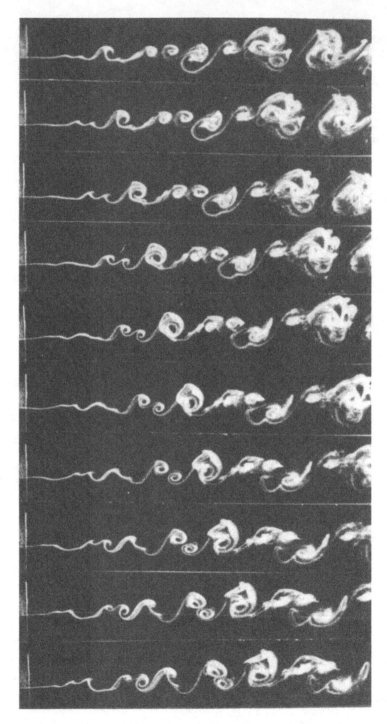

Plane mixing layer. Photograph by S. V. Sherikar, and R. Chevray. Reproduced from L. J. S. Bradbury, F. Durst, B. E. Launder, F. W. Schmidt, J. H. Whitelaw, *Turbulent Shear Flows 3*, Springer, New York, 1982.

3 Vorticity Laws

According to the flow characteristic under study, it is useful to emphasize either the role played by the pressure terms or that played by the velocity field acting on itself. The relationship between the pressure field and the velocity field is easily put in evidence for incompressible flows by considering both the continuity equation

$$\frac{\partial u_i}{\partial x_i} = 0 \qquad (3.1)$$

and the momentum equation

$$\frac{\partial u_i}{\partial t} + u_j \frac{\partial u_i}{\partial x_j} = -\frac{1}{\rho} \frac{\partial p}{\partial x_i} + f_i + \nu \frac{\partial^2 u_i}{\partial x_j \partial x_j} \qquad (3.2)$$

The divergence operator applied to Eq. 3.2 yields

$$\frac{\partial}{\partial x_i} \frac{\partial}{\partial t} u_i + \frac{\partial}{\partial x_i}\left(u_j \frac{\partial u_i}{\partial x_j}\right) = -\frac{1}{\rho} \frac{\partial^2 p}{\partial x_i \partial x_i} + \frac{\partial f_i}{\partial x_i} + \nu \frac{\partial}{\partial x_i}\left(\frac{\partial^2 u_i}{\partial x_j \partial x_j}\right)$$

and, taking into account the continuity equation, we find

$$\frac{1}{\rho} \frac{\partial^2 p}{\partial x_i \partial x_i} = \frac{\partial f_i}{\partial x_i} - \frac{\partial^2 u_i u_j}{\partial x_i \partial x_j} \qquad (3.3)$$

The vector $\mathbf{f}(f_i)$ that is the resultant of the volume forces per unit mass of fluid is here denoted by f. Whenever these forces are or can be considered negligible compared with the inertial forces, we have

$$\frac{1}{\rho} \frac{\partial^2 p}{\partial x_i \partial x_i} = -\frac{\partial^2 u_i u_j}{\partial x_i \partial x_j} \qquad (3.4)$$

The pressure field is related to the velocity field in a complex manner, and the laplacian operator expresses the nonlocal connection between the two. Viscous effects do not explicitly intervene in the relationship between pressure and velocity fields, yet they do alter the velocity distribution (and the boundary conditions for the pressure). Accordingly, the pressure field is seen to be altered by viscous effects through the velocity field only.

If we consider a uniformly loaded plate, the local strain is linked to the applied forces by the Laplace equation; this analogy illustrates the close interaction of the pressure field and the velocity field. Disregarding for a moment the role of the pressure field, we examine the role of the velocity field acting on itself. The momentum equation for inviscid flow is

$$\frac{\partial \mathbf{V}}{\partial t} + 2\mathbf{\Omega} \times \mathbf{V} = \text{grad}[\mathcal{H}] \tag{3.5}$$

in which the volume and pressure forces are embedded in the last term:

$$\mathcal{H} = \frac{V^2}{2} + U + h$$

in which $h = \int_0^p dp/\rho$ if the fluid is barotropic (density ρ is a function of pressure p only) and U is the potential energy per unit mass.

Taking the curl of Eq. 3.5 gives

$$\frac{\partial \text{ curl } \mathbf{V}}{\partial t} + 2 \text{ curl}(\mathbf{\Omega} \times \mathbf{V}) = 0$$

so we finally have

$$\frac{\partial \text{ curl } \mathbf{V}}{\partial t} + \text{curl}(\text{curl } \mathbf{V} \times \mathbf{V}) = 0$$

If we expand this last equation, we obtain the Helmholtz equations, which we present later in detail.

3.2 CROCCO'S THEOREM

A general relation between the sum $\mathcal{H}$ and the vorticity is readily written as

$$\frac{\partial \mathbf{V}}{\partial t} + 2\mathbf{\Omega} \times \mathbf{V} = \text{grad } \mathcal{H} \tag{3.6}$$

which is a general form of the Bernoulli theorem and is called Crocco's theorem.

For steady irrotational flows, the sum

$$\frac{V^2}{2} + \int \frac{dp}{\rho} + U$$

obviously can be only a function of time (this seems paradoxical, but for steady flow, only the velocity must be independent of time). This relationship, named after Daniel Bernoulli, was actually first derived by Euler.

For two-dimensional flows, Crocco's theorem can be interpreted as yielding the results presented in the figure:

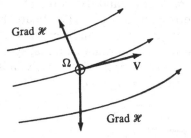

As we know, the Bernoulli equation applies along a streamline for steady flows. Crocco's theorem shows that, for irrotational flows, the equation applies throughout the flow. It also applies when vorticity and stream-lines coincide ($\Omega \times V = 0$), a case that is realized only under unsteady conditions (Beltrami–Gromeka's flows).

3.3 BASIC EQUATIONS

If we consider a small closed curve and draw the corresponding vortex lines through every point of this curve, we thus define a tube called a *vortex tube*. The fluid contained in such a tube constitutes a *vortex filament*, or simply *a vortex* (see figure).

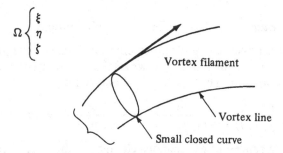

We can describe the velocity field in either an eulerian or a lagrangian system of coordinates. These systems are used whenever the properties of vortex filaments are considered. *Euleriaan coordinates* describe the field at every point in the laboratory frame as time passes. *Lagrangian coordinates* identify every particle and the motion of a particle is taken as a function of t (just as in particle mechanics). Having an eulerian description of the field enables us to infer a lagrangian description (even though this transformation is not easy to make), and vice versa.

3.3.a   *Lagrangian to eulerian system of coordinates*

Let x, y, z be the coordinates of a fixed point, thus no references are made to the flow field. In a lagrangian frame, the initial coordinates of a fluid particle are denoted a, b, c, or a_j, and its coordinates at time t are denoted x, y, z, or x_i; so a correspondence exists (see figure) between x, y, z and a, b, c that depends on time t and can be expressed as $x_i = x_i(a_j, t)$. Thus, a dependence is defined between an initial space (a, b, c; a_j) and a space defined at time t (x, y, z; x_j). Conversely, this point-to-point correspondence is given by $a_i = a_i(\mathbf{x}, t)$.

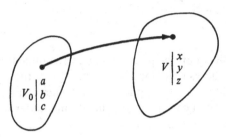

Two fluid elements cannot merge.

When x, y, z, denote the coordinates of a fluid particle at time t, the velocity of this fluid particle is given by $dx_i/dt = u_i(\mathbf{a}, t)$. If we substitute x, y, z, for a, b, c, we obtain

$$\frac{dx_i}{dt} = u_i(\mathbf{a}, t) = u_i(\mathbf{x}, t)$$

3.3.b   *Eulerian to lagrangian system of coordinates*

Reciprocally, it is possible to go from an eulerian frame to a lagrangian frame. Starting from an eulerian frame, we establish a connection through the streamline equation. The velocity field is given by

$$\frac{dx}{u(x, y, z, t)} = \frac{dy}{v(x, y, z, t)} = \frac{dz}{w(x, y, z, t)} = dt \qquad (3.7)$$

where dx, dy, dz represent a trajectory element of the fluid particle.

By integration we obtain $x_i = x_i(t, c_1, c_2, c_3)$, in which the three constants are determined from the initial conditions $x_i(t_0, c_1, c_2, c_3) = a_i$ and are given as functions of a_i. Substituting for c_1, c_2, and c_3 from the last equation in the previous one, we find that

$$x_i = x_i(\mathbf{a}, t)$$

This kind of formulation defines the trajectories of all the fluid particles. The history of any fluid element can then be easily followed. The second-order tensor $\partial x_i / \partial a_j$ plays a dominant role in the mechanics of continua, providing information about the evolution of particles that start at the same time t and are initially located next to each other. For instance, the first particle is located at point a_i and the second at point $a_i + \Delta a_i$. For any time t, we can write

$$\Delta x_i = \frac{\partial x_i}{\partial a_j} \Delta a_j$$

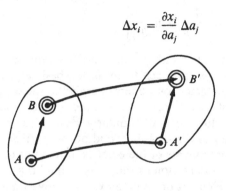

Although the two fluid elements under consideration must be close to each other, the two observations can be made at very different $t = 0$ and $t \neq 0$. This aspect of the problem must be emphasized. Regarding the motion of a fluid particle in an eulerian frame, we take a series expansion with respect to space variables only,

$$u_i(x + \delta x, y + \delta y, z + \delta z, t) = u_i(x, y, z, t) + \frac{\partial u_i}{\partial x}\delta x + \frac{\partial u_i}{\partial y}\delta y + \frac{\partial u_i}{\partial z}\delta z$$

so we go from **AB** to **A'B'** for time t. The two points under consideration must be close to each other, and the prediction is valid only for a small interval of time. In this very restrictive frame, the motion of the fluid particle is described by

$$\mathbf{V}'_B = \underset{\substack{\text{global} \\ \text{motion}}}{\mathbf{V}_A} + \underset{\text{rotation}}{\boldsymbol{\Omega} \times \mathbf{AB}} + \underset{\text{strain rate}}{\text{grad } f(\delta x, \delta y, \delta z)}$$

$$2f(\delta x, \delta y, \delta z) = a_1\,\delta x^2 + a_2\,\delta y^2 + a_3\,\delta z^2 + 2b_1\,\delta y\,\delta z$$

$$+ 2b_2\,\delta z\,\delta x + 2b_3\,\delta x\,\delta y$$

$$\begin{cases} a_1 = \dfrac{\partial u}{\partial x} \\[2mm] a_2 = \dfrac{\partial v}{\partial y} \\[2mm] a_3 = \dfrac{\partial w}{\partial z} \end{cases} \begin{cases} 2b_1 = \dfrac{\partial w}{\partial y} + \dfrac{\partial v}{\partial z} \\[2mm] 2b_2 = \dfrac{\partial u}{dz} + \dfrac{\partial w}{\partial x} \\[2mm] 2b_3 = \dfrac{\partial v}{dx} + \dfrac{\partial u}{\partial yp} \end{cases} \boldsymbol{\Omega} \begin{cases} 2\xi = \dfrac{\partial w}{\partial y} - \dfrac{\partial v}{\partial z} \\[2mm] 2\eta = \dfrac{\partial u}{\partial z} - \dfrac{\partial w}{\partial x} \\[2mm] 2\zeta = \dfrac{\partial v}{\partial x} - \dfrac{\partial u}{\partial y} \end{cases}$$

With this approach, the straining process is linked to the second-order tensor $\partial u_i/\partial x_j$. At time t the local rotation is described by the rotational operator. For a solid body, the rotation is the same at any point (at a given time). In a fluid, this rotation is a local property of the medium, and the operator giving the material derivative

$$\left\{ \frac{\partial}{\partial t} + u_i \frac{\partial}{\partial x_j} \right\}$$

must be considered to provide a link between the two representations.

We take advantage of this approach to recall the basic equations in a lagrangian frame. The velocity components parallel to the coordinate axis of the particle (a_i) at time t are $\partial x_i/\partial t$, and the components of the acceleration in the same directions are similarly $\partial^2 x_i/\partial t^2$. Let p be the pressure and ρ the density in the neighborhood of this particle at time t; X, Y, and Z are the components of the external forces per unit mass acting there. Considering the motion of the mass of fluid that at time t occupies the differential element of volume δx, δy, δz, we find that

$$\frac{\partial^2 x_i}{\partial t^2} = F_i - \frac{1}{\rho} \frac{\partial p}{\partial x_i}$$

These equations contain differential coefficients with respect to x_i whereas our independent variables are a_i, t. The basic variables are linked to the choice of a particle (lagrangian frame) and not to a choice of a point in space (eulerian frame). To eliminate these differential coefficients, we multiply the preceding equations first by $\partial x_i/\partial a$ and add, second by $\partial x_i/\partial b$ and add, and third by $\partial x_i/\partial c$ and add. We thus get the three equations

$$\left[\frac{\partial^2 x_j}{\partial t^2} - F_j \right] \frac{\partial x_j}{\partial a_i} + \frac{1}{\rho} \frac{\partial p}{\partial a_i} = 0 \tag{3.8}$$

To prevent confusion, the components of the external forces are called F_j in Eq. 3.8 (instead of a_j, which represents the initial coordinates of the particle). A lagrangian description establishes a point-to-point correspondence between the location of a fluid particle at time $t = 0$ and at an arbitrary time t. This correspondence can be extended to volumes, which allows us to write the continuity equation.

We consider an element of fluid that originally occupied a rectangular parallelepiped having its center at point (a, b, c) and its sides δa, δb, and δc parallel to the axes. At time t, the same element forms an oblique parallelepiped. The center now has the coordinates x, y, z. and the projections of the sides on the coordinate axes are, respectively,

$$\frac{\partial x_i}{\partial a_j} \delta a_j$$

The volume of the parallelepiped is therefore

$$\begin{vmatrix} \dfrac{\partial x}{\partial a} & \dfrac{\partial y}{\partial a} & \dfrac{\partial z}{\partial a} \\[2mm] \dfrac{\partial x}{\partial b} & \dfrac{\partial y}{\partial b} & \dfrac{\partial z}{\partial b} \\[2mm] \dfrac{\partial x}{\partial c} & \dfrac{\partial y}{\partial c} & \dfrac{\partial z}{\partial c} \end{vmatrix} \qquad \delta x\,\delta y\,\delta z = \frac{\partial(x,\,y,\,z)}{\partial(a,\,b,\,c)}\,\delta a\,\delta b\,\delta c$$

$$dV = J\,dV_0$$

The jacobian J of the point-to-point correspondence appears. It is closely related to the initial and final volume occupied by the fluid, so it becomes

$$\rho\,\frac{\partial(x,\,y,\,z)}{\partial(a,\,b,\,c)} = \rho_0 \qquad \rho\,dV = \rho_0\,dV_0$$

3.4   THE VORTICITY FLUX ACROSS A CLOSED SURFACE IS NULL

This is a direct consequence of the divergence theorem

$$\iint_S \mathbf{\Omega n}\,dS = \iiint_V \operatorname{div} \mathbf{\Omega}\,dV$$

where $\mathbf{n}$ is oriented outward from the volume V bounded by S. Because div curl $\mathbf{V} \equiv 0$, it follows that

$$\iint_S \mathbf{\Omega n}\,dS \equiv 0$$

In other words, because there is no vorticity flux across the boundary of a vortex filament, the vortex flow across a surface element of area dS and along the normal $\mathbf{n}$ of a vortex filament is constant:

$$|\mathbf{\Omega}_1|\,dS_1 = |\mathbf{\Omega}_2|\,dS_2$$

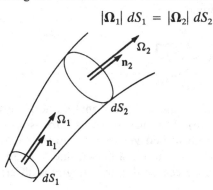

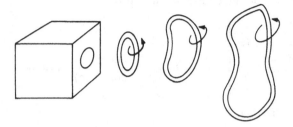

Fig. 3.1   Vortex rings issued from a hole

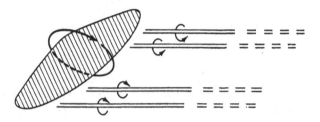

Fig. 3.2   Vortex lines from a wing

It is clear that isolated vortex lines cannot exist; they either form closed loops or end at a fluid boundary. Thus, ring-shaped vortex tubes may be observed in various cases (Fig. 3.1).

Vortex lines for a lifting wing can be drawn as in Fig. 3.2.

From the previous theorem, we see that free vortices must be associated with attached vortices that are linked to the lifting wing itself. The theorem is verified for the global pattern of vortices. Free vortices induce velocities at any station of the lifting wing. These induced velocities interact so that the resultant **v** (see figure) becomes normal to **V** at any station (**W** is a relative velocity). Even for an inviscid fluid, drag effects can be observed. They are directly connected to effects of the finite length of the wing.

In a two-dimensional field, fluid in irrotational motion exerts a side force but no drag (see figure). This side or lifting force is proportional to the fluid density, the undisturbed velocity **V** and the circulation Γ defined by $\Gamma = \oint_c \mathbf{V} \cdot \mathbf{t} \, dl$ taken along a closed line C (with element dl) surrounding the obstacle. Because the line integral is different from zero, the velocity field is disturbed; therefore, the pressure field is also distorted and generates the side force (see figure). The side force can be used to support an aircraft against gravity, or it can be used to generate axial momentum of the working fluid when the body is the blade of a rotating propeller or turbine.

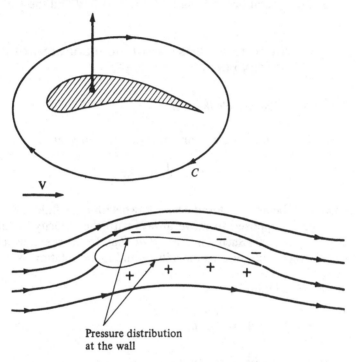

Pressure distribution
at the wall

Aeroplane wings and propeller blades are not infinitely long cylinders, so we must consider spanwise effects by taking into account vortex theory.

The line integral along C is closely related to the intensities of the vortex lines crossing any surface limited by C. We note that vortices are either concentrated at given points (indicated by stars in the figure) or

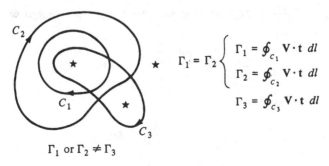

continuously distributed in space. In the first case a velocity potential can be introduced in an extended form; in the second case no possibility exists for introducing this concept.

For the case of continuously distributed vortices, the flow is rotational and the line integral depends on both the vector **V** and the path C.

3.5 THREE THEOREMS ON THE DEVELOPMENT OF VORTEX LINES

3.5.a *Lagrange's theorem*

Under limited conditions we can show that the circulation

$$\Gamma_c = \oint_C \mathbf{V} \cdot \delta\mathbf{l} \tag{3.9}$$

is constant, C being any closed curve drawn within the fluid; every point of this line is supposed to move always with the velocity of the fluid particle at the point. Such a line is a *fluid line*. In the previous integral, δl is not a vector because its magnitude and its orientation evolve with t.

For the circulation Γ,

$$\Gamma = \int_M^{M'} u_i \, \delta x_i$$

$$\frac{d\Gamma}{dt} = \int_M^{M'} \frac{du_i}{dt} \delta x_i + \int_M^{M'} u_i \frac{d}{dt} \delta x_i$$

If an acceleration potential Q exists, we have

$$\frac{du_i}{dt} = \frac{\partial Q}{\partial x_i}$$

Besides, a unique value of Q is supposed to be associated with each point

$$\int_M^{M'} \frac{du_i}{dt} \delta x_i = \int_M^{M'} \frac{\partial Q}{\partial x_i} \delta x_i = Q_{M'} - Q_M$$

On the other hand,

$$\frac{d}{dt} \delta x_i = \delta \frac{dx_i}{dt} = \delta u_i$$

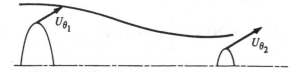

Fig. 3.3   Vortex tube contraction

The second integral can be evaluated:

$$\int_M^{M'} u_i \frac{d}{dt} \delta x_i = \int_M^{M'} u_i \, \delta u_i$$

$$= \frac{1}{2} [u^2 + v^2 + w^2]_M^{M'}$$

$$\frac{d}{dt} [\Gamma]_M^{M'} = Q_{M'} + \frac{1}{2} [u^2 + v^2 + w^2]\Big|_{M'}$$

$$- Q_M - \frac{1}{2} [u^2 + v^2 + w^2]\Big|_{M}$$

The time evolution of the integral evaluated along a material line is measured by the difference of the term $Q + \frac{1}{2}V^2$ at points M and M' at time t. If the line integral is taken along a closed path and if the acceleration potential has a unique value at each point, then

$$\frac{d\Gamma_C}{dt} = 0 \qquad (3.10)$$

(a unique value of the velocity can be defined at a given point, and a unique fluid particle is located at a point).

In the flow under consideration the line integral is constant; if this line integral is null at the initial time, it is null at any subsequent time. We introduce a velocity potential at this stage because we know how important the simplification introduced by this concept can be. A simple example is presented in Fig. 3.3, which shows that the tangential velocity U_θ experiences a large increase after going through a contraction of the vortex tube.

3.5.b   *Helmholtz's theorem*

The evolution of the vorticity can be predicted by a differential equation. We start with the equation

$$\frac{\partial \mathbf{V}}{\partial t} + 2\mathbf{\Omega} \wedge \mathbf{V} = \text{grad } \mathcal{H} \qquad (3.11)$$

which can be expressed as

$$\text{curl}\left[\frac{\partial \mathbf{V}}{\partial t} + 2\mathbf{\Omega} \wedge \mathbf{V}\right] = \text{curl grad } \mathcal{H} = 0 \qquad (3.12)$$

after we apply the curl operator to the momentum equation. The body forces are assumed to be imbedded in a potential function, and a relationship is assumed between p and ρ. This operation eliminates $\mathcal{H}$, and the resulting equation shows clearly the interaction between velocity and curl fields. The components of the local vorticity are termed ξ, η, and ζ, so the previous equation can be expanded as

$$\left\{ \begin{array}{l} \dfrac{\partial u}{\partial t} + 2\{\eta w - \zeta v\} = \dfrac{\partial \mathcal{H}}{\partial x} \\[2mm] \dfrac{\partial v}{\partial t} + 2\{\zeta u - \xi w\} = \dfrac{\partial \mathcal{H}}{\partial y} \\[2mm] \dfrac{\partial w}{\partial t} + 2\{\xi v - \eta u\} = \dfrac{\partial \mathcal{H}}{\partial z} \end{array} \right\}$$

The first component of Eq. 3.12 is

$$\frac{\partial}{\partial t}\left\{\frac{\partial v}{\partial z} - \frac{\partial w}{\partial y}\right\} + 2\left\{\frac{\partial}{\partial z}(\zeta u - \xi w) - \frac{\partial}{\partial y}(\xi v - \eta u)\right\} = 0$$

which can be expressed as

$$\frac{\partial \xi}{\partial t} - u\left\{\frac{\partial \eta}{\partial y} + \frac{\partial \zeta}{\partial z}\right\} + v\frac{\partial \xi}{\partial y} + w\frac{\partial \xi}{\partial z}$$

$$+ \xi\left\{\frac{\partial v}{\partial y} + \frac{\partial w}{\partial z}\right\} - \zeta\frac{\partial u}{\partial z} - \eta\frac{\partial u}{\partial y} = 0 \qquad (3.13)$$

Rearrangements can be made by using both the continuity equation

$$\frac{1}{\rho}\frac{d\rho}{dt} + \frac{\partial u}{\partial x} + \frac{\partial v}{\partial y} + \frac{\partial w}{\partial z} = 0$$

and the identity

$$\text{div } \mathbf{\Omega} = \frac{\partial \xi}{\partial x} + \frac{\partial \eta}{\partial y} + \frac{\partial \zeta}{\partial z} = 0$$

Taking this last relation into account makes the advective motion of ξ apparent, so the first five terms of Eq. 3.13 give the following evolution equation:

$$\frac{\partial \xi}{\partial t} + u\frac{\partial \xi}{\partial x} + v\frac{\partial \zeta}{\partial y} + w\frac{\partial \xi}{\partial z} = \frac{d\xi}{dt}$$

In the same equation the last four terms become

$$\xi\left(\frac{\partial v}{\partial y} + \frac{\partial w}{\partial z}\right) - \zeta\frac{\partial u}{\partial z} - \eta\frac{\partial u}{\partial y} = -\xi\left(\frac{\partial u}{\partial x} + \frac{1}{\rho}\frac{dp}{dt}\right) - \zeta\frac{\partial u}{\partial z} - \eta\frac{\partial u}{\partial y}$$

Final rearrangements yield

$$\frac{d\xi}{dt} - \frac{\xi}{\rho}\frac{d\rho}{dt}\frac{\partial\xi}{\partial x} = \xi\frac{\partial u}{\partial x} + \eta\frac{\partial u}{\partial y} + \zeta\frac{\partial u}{\partial z}$$

$$\frac{d}{dt}\left(\frac{\xi}{\rho}\right) = \frac{\xi}{\rho}\frac{\partial u}{\partial x} + \frac{\eta}{\rho}\frac{\partial u}{\partial y} + \frac{\zeta}{\rho}\frac{\partial u}{\partial z}$$

Similar equations can be derived for the other two components. In compressed form we can write

$$\frac{d}{dt}\left(\frac{\xi_i}{\rho}\right) = \frac{\xi_j}{\rho}\frac{\partial u_i}{\partial x_j}$$

Because a close connection exists between the velocity and the vorticity fields, the last equation expresses a nonlinear process. The vortex filaments are seen to be either convected by the velocity field (left-hand side) or distorted by this same velocity through gradient terms (right-hand side). A part of the nonlinear mechanisms imbedded in $\mathcal{H}$ has disappeared through the algebra. It can now be seen that the interaction between the velocity and the vortex fields is directly connected to linear mechanisms; this interaction generally exists, except for special flow fields (e.g., two-dimensional fields). If the vorticity is initially null everywhere, $\xi_0 = \eta_0 = \zeta_0 = 0$, then it remains null at any time t such as $t > t_0$. In other words, if the initial motion admits a velocity potential, this potential field exists at all times.

Note at this point that the foregoing analysis can be applied to inviscid fluids only. It is very simple to extend this kind of analysis to Newtonian fluids because in that case no nonlinear effects are introduced through the rheological behavior. For an incompressible fluid we find that

$$\underbrace{\frac{d\xi_i}{dt}}_{\text{advective terms}} = \underbrace{\xi_j\frac{\partial u_i}{\partial x_i}}_{\substack{\text{interaction between vorticity} \\ \text{and velocity fields}}} + \underbrace{\nu\frac{\partial^2\xi_i}{\partial x_j\partial x_j}}_{\text{viscous effects}}$$

Hence vorticity can be generated by viscous effects.

3.5.c   *Cauchy's theorem*

Cauchy's theorem can be considered as an integral form of Helmholtz's theorem. If the fluid motion is described in a lagrangian frame, the vortex field at any time can be deduced from the initial vortex field without integrating.

We start with the equation

$$\frac{d}{t}\left(\frac{\xi}{\rho}\right) = \frac{\xi}{\rho}\frac{\partial u}{\partial x} + \frac{\eta}{\rho}\frac{\partial u}{\partial y} + \frac{\zeta}{\rho}\frac{\partial u}{\partial z}$$

which can be written

$$\frac{d}{dt}\left(\frac{\xi}{\rho}\right) = \frac{|\Omega|}{\rho}\left\{\frac{\xi}{|\Omega|}\frac{\partial u}{\partial x} + \frac{\eta}{|\Omega|}\frac{\partial u}{\partial y} + \frac{\zeta}{|\Omega|}\frac{\partial u}{\partial z}\right\}$$

where $\xi/|\Omega|$, $\eta/|\Omega|$, and $\zeta/|\Omega|$ are the direction cosines of the vorticity vector Ω (α, β, γ). $|\delta l|$ stands for the modulus of a differential element of a vortex line, and δV represents the variation of V along such a vortex line. The components of δV are δu, δv, and δw. Then

$$\frac{\delta u}{|dl|} = \frac{\partial u}{\partial x}\frac{\delta x}{|\delta l|} + \frac{\partial u}{\partial y}\frac{\delta y}{|\delta l|} + \frac{\partial u}{\partial z}\frac{\delta z}{|\delta l|}$$

The direction cosines appear again, hence we can substitute $\delta u/|\delta l|$ for

$$\left\{\frac{\xi}{|\Omega|}\frac{\partial u}{\partial x} + \frac{\eta}{|\Omega|}\frac{\partial u}{\partial y} + \frac{\zeta}{|\Omega|}\frac{\partial u}{\partial z}\right\}$$

and get the very compressed expression

$$\frac{d}{dt}\left(\frac{\xi}{\rho}\right) = \frac{|\Omega|}{\rho}\frac{\delta u}{\delta l}$$

which corresponds to the vectorial form

$$\frac{d}{dt}\left(\frac{\Omega}{\rho}\right) = \frac{|\Omega|}{\rho}\frac{\delta V}{|\delta l|}$$

It is possible to show that $\delta V/|\delta l|$ is directly related to the stretching process of the vortex lines.

Let us consider an element of a vortex filament and the two points M and M' (see figure).

$$MM' = \delta l$$

$$V_M = \frac{d}{dt}OM \qquad V_{M'} = \frac{d}{dt}OM'$$

$$\delta V = V_{M'} - V_M = \frac{d}{dt}[OM' - OM] = \frac{d}{dt}\delta l$$

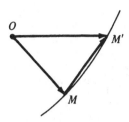

We can easily evaluate $\delta\mathbf{l}$ if the fluid motion is defined in a lagrangian frame:

$$\frac{d}{dt}\frac{\boldsymbol{\Omega}}{\rho} = \frac{|\boldsymbol{\Omega}|}{\rho|\delta\mathbf{l}|}\frac{d}{dt}(\delta\mathbf{l})$$

In passing notice that (see figure)

$$\frac{|\boldsymbol{\Omega}|}{\rho|\delta\mathbf{l}|} = \frac{|\boldsymbol{\Omega}_0|}{\rho_0|\delta\mathbf{l}_0|} = \text{constant}$$

This is a direct consequence of two theorems established in connection with vorticity mechanisms. In a lagrangian frame, the continuity equation gives

$$\rho\,\delta S|\delta\mathbf{l}| = \text{constant}$$

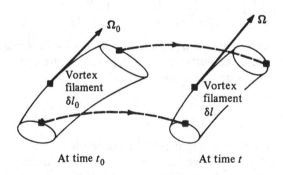

At time t_0 At time t

and this information is supplemented by the flux theorem

$$\delta S|\boldsymbol{\Omega}| = \text{constant}$$

so that

$$\frac{\rho|\delta\mathbf{l}|}{|\boldsymbol{\Omega}|} = \text{constant}$$

Finally we find that

$$\frac{d}{dt}\left[\frac{\Omega}{\rho}\right] = \frac{|\Omega_0|}{\rho_0|\delta l_0|}\frac{d}{dt}[\delta l]$$

$$\Delta_t\left[\frac{\Omega}{\rho}\right] = \frac{|\Omega_0|}{\rho_0|\delta l_0|}\Delta_t[\delta l] = \frac{|\Omega_0|}{\rho_0|\delta l_0|}[\delta l_t - \delta l_0]$$

the *x*-component of which is

$$\Delta_t\left[\frac{\xi}{\rho}\right] = \frac{|\Omega_0|}{\rho_0|\delta l_0|}[\delta x_t - \delta x_0]$$

The motion of a fluid particle is defined by

$$x_i = x_i(a_j, t) \qquad \text{or} \qquad \mathbf{x} = \mathbf{x}(a, b, c, t)$$

$$\Delta_t\left[\frac{\xi}{\rho}\right] = \frac{|\Omega_0|}{\rho_0|\delta l_0|}\left[\frac{\partial x}{\partial a}\delta a + \frac{\partial x}{\partial b}\delta b + \frac{\partial x}{\partial c}\delta c\right]$$

$$= \frac{|\Omega_0|}{\rho_0}\left[\frac{\partial x}{\partial a}\frac{\delta a}{|\delta l_0|} - \frac{\partial x}{\partial b}\frac{\delta b}{|\delta l_0|} + \frac{\partial x}{\partial c}\frac{\delta c}{|\delta l_0|}\right]$$

$$\frac{\delta a}{|\delta l_0|} = \alpha_0 \qquad \frac{\delta b}{|\delta l_0|} = \beta_0 \qquad \frac{\delta c}{|\delta l_0|} = \gamma_0$$

$$\Delta_t\left[\frac{\xi}{\rho}\right] = \left[\frac{\xi_0}{\rho_0}\frac{\partial x}{\partial a} + \frac{\eta_0}{\rho_0}\frac{\partial x}{\partial b} + \frac{\xi_0}{\rho_0}\frac{\partial x}{\partial c}\right]_{\Delta t}$$

The value of the derivatives $\partial x/\partial a$, $\partial x/\partial b$, and $\partial x/\partial c$ must be considered at time t_0 and t because the vortex motion depends on the evolution of the vortex line during a time interval $t - t_0$.

$$\Delta_t\left[\frac{\xi_i}{\rho}\right] = \frac{\xi_{j0}}{\rho_0}\frac{\partial x_i}{\partial x_j}$$

The lagrangian frame is rarely used, but the result of using it is fascinating. The evolution of the vorticity is thereby linked to the relative position of the two points of a vortex line. The role played by the initial conditions (ξ_0, η_0, ζ_0) is evident, and the concept of velocity potential is consistent with the equation of motion as far as inviscid flows are concerned. The remark that "the development of a vortex line is linked to a pure problem of geometry" has to be handled with caution because it is a consequence of the starting equation, which supposes the motion to be defined with respect to a galilean frame. With a rotating frame, the equations must be supplemented by inertial terms, and the conclusion is modified. The example of flow through a convergent duct, treated

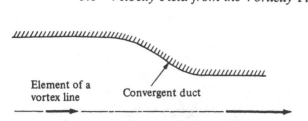

Fig. 3.4   Straining of a vortex line by a convergent duct

previously, can also be solved by using this last equation, as shown in Fig. 3.4.

3.6   THE VELOCITY FIELD DEDUCED FROM THE VORTICITY FIELD

Without detailing the most general correspondence between the velocity and vorticity fields, we want to show the kind of connection that can be established. This connection is very useful in describing electric fields, for instance. In fact, the conductors are the vortex lines of the magnetic field. Accordingly, the singular lines of the magnetic field are predetermined. The strength of the magnetic field is deduced from the intensities of these vortex lines, and a complex magnetic potential can therefore be defined. In fluid mechanics, the velocity field is usually given; hence, the vorticity distribution is extracted from the velocity field. However, in some cases, an induced velocity field, known from its vortex lines, is superimposed on the basic field. Such a situation is encountered in many industrial problems in which lift effects play a predominant role. We illustrate with a vortex scheme for a two-bladed propeller. The basic field results from the velocity of the aircraft engine, which is more or less altered by the shapes of the bodies existing in the vicinity of the propeller. The vortex lines are roughly shown in Fig. 3.5.

3.6.a   *A unicity theorem*

Let's suppose that two velocity fields V and V' correspond to the same vorticity field Ω at a given time:

$$V(x, y, z) \qquad V'(x, y, z)$$

$$\Omega(x, y, z)$$

Accordingly we have

$$\left. \begin{array}{l} \Omega = \text{curl } V \\ \Omega = \text{curl } V' \end{array} \right\} \ \forall \, x \in D$$

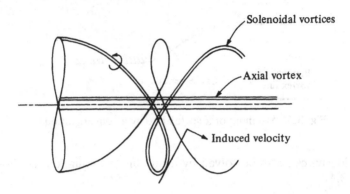

Fig. 3.5   Velocity induced by a system having two solenoidal vortices and an axial vortex

This leads to curl$(V - V') = 0$, which implies that $V - V' =$ grad Φ, in which Φ is a scalar potential function. When dealing with an incompressible fluid we have

$$\text{div } V = 0 \quad \text{and} \quad \text{div } V' = 0$$

so

$$\text{div}(V - V') = 0$$

As for the potential ϕ associated with the vectorial field $V - V'$, we have

$$\text{div grad } \Phi = \Delta\Phi = 0$$

The fluid is supposed to be enclosed in a domain D inside a surface S. That becomes

$$V \cdot n = 0 \quad V' \cdot n = 0$$

$$(V - V') \cdot n = 0 \Rightarrow \frac{\partial\Phi}{\partial n} = 0$$

Therefore, Φ is a harmonic function in the whole domain D limited by S; moreover, $\partial\Phi/\partial n = 0$; $\forall\, M \in S$. These conditions require Φ to be a constant, which may be set equal to zero, and therefore $V \equiv V'$. The velocity field corresponding to a given vorticity field is unique.

3.6.b   *The connection between the vorticity field and the velocity field*

In this case the vorticity field is supposed to be known. Using the nabla vector ∇ for the sake of brevity, we introduce a vectorial Poisson's equation whose solution is easy to determine.

Given $\boldsymbol{\Omega}(\mathbf{x})$, we have

$$\nabla \cdot \boldsymbol{\Omega} = 0 \quad \text{and} \quad \nabla \times \mathbf{V} = \boldsymbol{\Omega}(\mathbf{x}) \quad \text{(given field)}$$

We next suppose the unknown vectorial field to be solenoidal, so

$$\text{div } \mathbf{V} = 0 \Leftrightarrow \nabla \cdot \mathbf{V} = 0$$

The unknown field being solenoidal, we can introduce a vector potential **B** such that

$$\mathbf{V} = \nabla \times \mathbf{B} \tag{3.14}$$

At this stage **B** is assumed to have only first and second derivatives. We write

$$\nabla \times \mathbf{V} = \nabla \times \nabla \times \mathbf{B} = \boldsymbol{\Omega}$$

and introduce the relation

$$\nabla \times \nabla \times \mathbf{B} = \nabla[\nabla \cdot \mathbf{B}] - \nabla^2 \mathbf{B} = \boldsymbol{\Omega}$$

We still suppose

$$\nabla \cdot \mathbf{B} = 0 \Leftrightarrow \text{div } \mathbf{B} = 0$$

Disregarding at this stage the consequences of this hypothesis on the solution, we write

$$-\nabla^2 \mathbf{B} = \boldsymbol{\Omega} \tag{3.15}$$

in which **B** is defined by a vectorial Poisson's equation. The general solution of this equation is composed of the solution of the homogeneous equation

$$-\nabla^2 \mathbf{B} = 0 \tag{3.16}$$

which takes into account the vortex elements located outside the surface S, supplemented by a particular solution of the complete equation, which takes into account the vortex elements located inside the surface S (i.e., in the domain D).

In other words, the role played by the vortex elements located outside S is entirely determined by the boundary conditions given at any point $M(x)$ belonging to S. This mathematical splitting of the general solution into the homogeneous equation and the particular solution of the complete equation can always be interpreted physically. It is remarkable that external disturbances can be taken into account through boundary conditions only. Accordingly, if the domain D is large enough to include all the vortex elements, the particular solution of the complete equation should be adequate to treat our problem. A point $M'(X')$ where the vortex element is located can be identified, whereas the effect of this disturbance is observed at point $M(X)$. This leads to

$$\mathbf{B(x)} = \frac{1}{4\pi} \int_0 \frac{\mathbf{\Omega'(x')}}{|\mathbf{r}|} d(\mathbf{x'})$$

$$\mathbf{r} = \mathbf{x'} - \mathbf{x}$$

and $d(x')$ stands for a volume element. This solution is available if we also have

$$\nabla_x \cdot \mathbf{B} = 0$$

(the divergence of $\mathbf{B}$ has been supposed to be null). Accordingly we must have

$$\nabla_x = \left\{ \frac{1}{4\pi} \int_0 \frac{\mathbf{\Omega'(x')}}{|\mathbf{r}|} d(\mathbf{x'}) \right\} \tag{3.17}$$

Noted that the operator has an antisymmetrical role when applied to a Cauchy kernal $1/|\mathbf{r}|$. We make some rearrangements starting from this development:

$$\frac{1}{4\pi} \int_0 \nabla_x \cdot \left\{ \frac{\mathbf{\Omega'}}{|\mathbf{r}|} \right\} d(\mathbf{x'}) = 0$$

$$\frac{1}{4\pi} \int_D \mathbf{\Omega'} \, \nabla_x \cdot \left\{ \frac{1}{|\mathbf{r}|} \right\} d(\mathbf{x'}) = \frac{1}{4\pi} \int_D \frac{1}{|\mathbf{r}|} \nabla_x \cdot \mathbf{\Omega'} \, d(\mathbf{x'}) = 0$$

$$-\frac{1}{4\pi} \int_D \mathbf{\Omega'} \, \nabla_x \left\{ \frac{1}{|\mathbf{r}|} \right\} d(\mathbf{x'}) + \frac{1}{4\pi} \int_D \frac{1}{|\mathbf{r}|} \nabla_{x'} \cdot \mathbf{\Omega'} \, d(\mathbf{x'}) = 0$$

The requisite condition to validate this computation can be written

$$\nabla_x \frac{1}{4\pi} \int_0 \frac{\mathbf{\Omega'}}{|\mathbf{r}|} d(\mathbf{x}) = \frac{1}{4\pi} \int_D \mathbf{\Omega'} \nabla_x \left\{ \frac{1}{\bar{r}} \right\} d(\mathbf{x'})$$

$$= -\frac{1}{4\pi} \int_D \nabla_x \left\{ \frac{\mathbf{\Omega}}{|\mathbf{r}|} \right\} d(\mathbf{x}) + \frac{1}{4\pi} \int_D \frac{1}{|\mathbf{r}|} \nabla_{x'} \cdot \mathbf{\Omega'} d(\mathbf{x'})$$

The last integral is, of course, null:

$$\nabla_{x'} \mathbf{\Omega'} = 0 \Leftrightarrow \operatorname{div} \mathbf{\Omega'} = 0$$

As for the first integral, the right-hand side of the equation can be transformed to emphasize the unique role of the boundary conditions by the divergence theorem. That becomes

$$-\frac{1}{4\pi} \int_D \nabla_{x'} \cdot \left\{ \frac{\mathbf{\Omega'}}{|\mathbf{r}|} \right\} d(\mathbf{x'}) - \frac{1}{4\pi} \int_{S'} \frac{\mathbf{\Omega'} \cdot \mathbf{n'}}{|\mathbf{r}|} dS'$$

which yields

$$\frac{1}{4\pi} \int_D \Omega \nabla_{x'} \left\{ \frac{1}{|\mathbf{r}|} \right\} d(\mathbf{x}') = \frac{1}{4\pi} \int_{S'} \frac{\Omega' \cdot \mathbf{n}'}{|\mathbf{r}|} dS'$$

The additional condition imposed on **B** is verified if the flux integral of Ω' over a surface S' is null, which is possible in the two cases: if Ω is null outside of D and is also null at any point of S'; or if at any point of S'; Ω' is tangent to the surface. Taking into account these conditions, we write

$$\mathbf{B}(\mathbf{x}) = \frac{1}{4\pi} \int_D \frac{\Omega'(\mathbf{x}')}{|\mathbf{r}|} d(\mathbf{x}') \tag{3.18}$$

and

$$\mathbf{V} = \nabla_{\mathbf{x}} \times \mathbf{B}$$

$$\mathbf{V} = \left(\frac{1}{4\pi} \int \frac{\Omega'(\mathbf{x}')}{|\mathbf{r}|} d(\mathbf{x}') \right) \nabla_{\mathbf{x}}$$

Since Ω', which is located at point $\mathbf{x}'$, is not affected by the operator acting at point $\mathbf{x}$, it is sufficient to take the curl of the Cauchy kernel $1/|\mathbf{r}|$

$$\Omega \begin{cases} \xi' \\ \eta' \\ \zeta' \end{cases} \qquad \mathbf{r} \begin{cases} x' - x \\ y' - y \\ z' - z \end{cases}$$

which becomes

$$- \frac{(x' - x)\eta' + (y' - y)\xi'}{|\mathbf{r}|^3}$$

Finally we write

$$\mathbf{V} = \nabla_x \times \mathbf{B} = -\frac{1}{4\pi} \int \frac{\mathbf{r} \times \Omega'}{|\mathbf{r}|^3} d(\mathbf{x}')$$

$$\mathbf{V} = -\frac{1}{4\pi} \int \frac{\mathbf{r} \times \Omega'}{|\mathbf{r}|^3} d(\mathbf{x}')$$

This result can also be presented in the form

$$d\mathbf{V}(P) = \frac{\Gamma}{4\pi} \frac{\mathbf{PM} \times d\mathbf{M}}{(|\mathbf{PM}|)^3}$$

The influence of an elementary vortex $d\mathbf{M}$, fictitiously isolated, is observed at point P.

3.7 IMPLICATIONS OF THE VORTEX THEORY IN COMPLEX COMPUTATIONS

3.7.a *The motion of an incompressible fluid*

is governed by the momentum equation

$$\frac{\partial \mathbf{V}}{\partial t} + [\mathbf{V} \cdot \mathbf{\nabla}]\mathbf{V} = -\frac{1}{\rho}\mathbf{\nabla}p - \nu\mathbf{\nabla}^2\mathbf{V} \tag{3.19}$$

supplemented by the continuity equation

$$\mathbf{\nabla} \cdot \mathbf{V} = 0 \tag{3.20}$$

the solution of which must satisfy both boundary and initial conditions.

At least two fundamental difficulties appear in solving these equations. First, the momentum equation can be considered as generating a first set of functions; in other words, the solution of this equation belongs to a *functional space*. Taking into account the continuity equation, we must work in a subset of this initial functional space; in fact, the continuity equation appears to be a nontrivial additional condition. The second difficulty is associated with the role of the pressure, which appears to be a typical parameter that is not governed by an equation including the usual advective, production, and diffusive terms. It is in connection with the whole flow (except for particular cases) that when the vorticity of the flow is null, a simple relationship exists between the pressure and the velocity fields. This case is thoroughly discussed in elementary courses.

The continuity equation for two-dimensional flows can be verified simply by introducing a stream function $\Psi(x, y)$, defined so that

$$u = \frac{\partial \Psi}{\partial y} \qquad v = -\frac{\partial \Psi}{\partial x}$$

The lines corresponding to $\Psi(x, y)$ = constant are obviously streamlines, and the unique component of the vorticity ζ is governed by

$$\frac{\partial \zeta}{\partial t} + u\frac{\partial \zeta}{\partial x} + v\frac{\partial \zeta}{\partial y} = \nu\left\{\frac{\partial^2 \zeta}{\partial x^2} + \frac{\partial^3 \zeta}{\partial y^2}\right\}$$

Since

$$2\zeta = \frac{\partial v}{\partial x} - \frac{\partial u}{\partial y} = -\frac{\partial^2 \Psi}{\partial x^2} - \frac{\partial^2 \Psi}{\partial y^2}$$

or

$$-\nabla\Psi = 2\zeta \tag{3.21}$$

the initial problem degenerates into a Poisson's problem. The boundary conditions can be expected to open a more or less complex problem, introducing a limiting value for ζ that corresponds to the nonslip condition.

This approach produces two significant simplifications: The difficulty introduced by the continuity equation is removed, and the pressure term does not now explicitly appear in the new problem. We might therefore expect to find an efficient method for three-dimensional flows based on similar ideas, but unfortunately, introducing a stream function is much more difficult.

We have shown previously that a vectorial field is dominated by two kinds of singularities that affect the values of the two vectorial operators

$$\text{div } \mathbf{V} \quad \text{or} \quad \text{curl } \mathbf{V}$$

In fact, every complex vectorial field can be split into two parts: a vectorial field $\mathbf{V}_1$ with div $\mathbf{V}_1 = 0$ (solenoidal field) and an irrotational field $\mathbf{V}_2$ such as curl $\mathbf{V}_2 = 0$.

The vectorial field $\mathbf{V}_1$ is readily associated with any vectorial field through the curl operator:

$$\mathbf{V}_1 = \nabla \times \mathbf{\Psi} \Rightarrow \text{div } \mathbf{V}_1 = 0$$

with $\mathbf{\Omega} = \nabla \times \mathbf{V}_1$, which leads to $\nabla \times \nabla \times \mathbf{\Psi} = \mathbf{\Omega}$. The irrotational field V_2 can be expressed as

$$\mathbf{V}_2^2 = \nabla \Phi$$

which finally leads to $\mathbf{V} = \mathbf{V}_1 + \mathbf{V}_2 = \nabla \Phi + \nabla \times \mathbf{\Psi}$. From the term $\nabla \times \nabla \times \mathbf{\Psi}$ a vectorial Poisson equation appears for $\mathbf{\Psi}$, and a scalar potential equation is simultaneously introduced.

The treatment of these two equations is closely coupled. Ω is influenced by the tensorial velocity gradient through the Helmoltz equation, and this gradient for the whole flow V is equal to

$$\frac{\partial u_i}{\partial x_j} = \frac{1}{2}\left(\frac{\partial u_i}{\partial x_j} + \frac{\partial u_j}{\partial x_i}\right) + \frac{1}{2}\left(\frac{\partial u_i}{\partial x_j} - \frac{\partial u_j}{\partial x_i}\right) \tag{3.22}$$

Taking this splitting into account, the Helmoltz equation becomes

$$\frac{\partial \mathbf{\Omega}}{\partial t} + [\mathbf{V} \cdot \nabla]\mathbf{\Omega} = [\mathbf{\Omega} \cdot \nabla]\mathbf{V} + \nu \nabla \mathbf{\Omega}$$

$$\frac{\partial \mathbf{\Omega}}{\partial t} + [(\nabla \Phi + \nabla \times \mathbf{\Psi}) \cdot \nabla]\mathbf{\Omega} - [\mathbf{\Omega} \cdot \mathbf{V}](\nabla \Phi + \nabla \times \mathbf{\Psi}) = \nu \nabla \mathbf{\Omega}$$

The strong interaction dependence between the two fields clearly appears through this equation; it can be reduced if we assume that the second derivative with respect to Ω is not of the same order of magnitude in all directions–for instance, by making the following hypothesis (see figure);

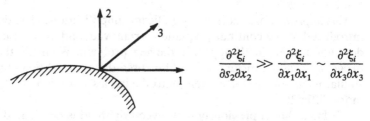

$$\frac{\partial^2 \xi_i}{\partial s_2 \partial x_2} \gg \frac{\partial^2 \xi_i}{\partial x_1 \partial x_1} \sim \frac{\partial^2 \xi_i}{\partial x_3 \partial x_3}$$

The results are similar to those introduced in the boundary layer theory and which lead to parabolic equations. The evolution of ξ is given by the previous equation. Moreover, the splitting of the boundary conditions between the two fields V_1 and V_2 is questionable. The initial conditions relative to Ω are easily introduced. Moreover, the problem created by the pressure term is overcome. At each stage of the computations the potential part of the velocity field gives the compact character of the field. The computations are rather difficult, but the convergence of the numerical process does not pose a special problem. We now sketch the method.

$$
\mathbf{V} \Rightarrow \\
\mathrm{div}\ \mathbf{V} = 0
\begin{cases}
\mathbf{V}_1 \begin{cases}
\mathbf{V}_1 = \mathbf{\nabla} \times \mathbf{\Psi} \\
\Downarrow \\
\mathrm{div}\ \mathbf{V}_1 = 0
\end{cases} \Rightarrow \mathbf{\nabla} \times \mathbf{\nabla\Psi} = \mathbf{\Omega} \ \text{Poisson's equation for } \Psi \\[2ex]
\mathbf{V}_2 \begin{cases}
\mathrm{div}\ \mathbf{V}_2 = 0 \quad\quad \mathbf{V}_2 = \mathrm{grad}\ \Phi \Rightarrow \Delta\Phi = 0 \\
\quad\quad\quad\quad\quad\quad\quad\quad \Downarrow \\
\quad\quad\quad\quad\quad\quad \mathbf{\nabla} \times \mathbf{V}_2 = 0
\end{cases}
\end{cases}
$$

$$\frac{\partial \xi_i}{\partial t} + V_j \frac{\partial \xi_i}{\partial x_j} = \xi_j \frac{\partial v}{\partial x_j} + v \frac{\partial^2 \xi_i}{\partial x_j \partial x_j} \tag{3.23}$$

The last term of this equation is approximated by

$$\frac{\partial^2 \xi_i}{\partial x_l \partial x_l}$$

Remark: To display some interdependence between the velocity and vortex fields, two examples are presented in Figs. 3.6 and 3.7.

3.7.b

This example is directly related to the problem of a compressible turbulent field. The computations do not require any special knowledge in

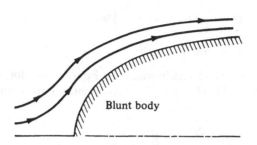

Fig. 3.6   The straining induced by a blunt body

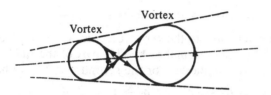

Fig. 3.7   The straining process induced by two rotating vortices

this domain, and the problem is analyzed using the aforementioned splitting:

$$\mathbf{V}(\mathbf{x}, t) = \text{curl}[\mathbf{\Psi}(\mathbf{x}, t)] + \text{grad}[\Phi(\mathbf{x}, t)]$$

incompressible part of the fluid   compressible part of the fluid

We set

$$\mathbf{V}_1 = \text{curl }\mathbf{\Psi} \quad \text{and} \quad \mathbf{V}_2 = \text{grad }\Phi$$

Here, our interest is focused on the dissipative processes, and our purpose consists of assessing the role of the viscous forces on each of the compressible and the incompressible modes. We first introduce a new function C defined as

$$C = \text{div }\mathbf{V} = \text{div }\mathbf{V}_2 = \text{div}(\text{grad }\Phi) = \Delta\Phi$$

On the other hand, we can write

$$\text{curl }\mathbf{V} = \text{curl}(\text{curl }\mathbf{\Psi}) = \text{grad}(\text{div }\mathbf{\Psi}) - \Delta\mathbf{\Psi}$$

Except for two-dimensional fields in which $\mathbf{\Psi}$ has a unique component that can be identified with the stream function, we must consider $\mathbf{\Psi}$ to be a vector. For three-dimensional flows, this vector should have only two components but we prefer to consider $\mathbf{\Psi}$ a complete vector (with three components) and impose on it the additional condition div $\mathbf{\Psi} = 0$, which enables us to write equations for $\mathbf{\Psi}$ and Φ that are similar:

$$\Omega = \frac{1}{2} \operatorname{curl} \mathbf{V} = -\frac{1}{2} \Delta \Psi$$

$$C = \operatorname{div} \mathbf{V} = \Delta \Phi$$

These two modes are easy to identify when the Fourier transforms are applied to the velocity field. The two previous equations are written in Fourier space as

$$\hat{\Psi} = \frac{2\hat{\Omega}}{k^2} \qquad \hat{\Phi} = \frac{1}{k^2} \hat{C}$$

Moreover,

$$\operatorname{div} \mathbf{V} = 0 \Rightarrow k_j \hat{u}_j^1 = 0$$

so the vector $\mathbf{V}_1$ is orthogonal to the wave vector $\mathbf{k}$, which requires this component to be in a plane normal to $\mathbf{k}$. This part of the fluid corresponds to an incompressible mode. For the second part $\mathbf{V}_2$, we have

$$\operatorname{curl} \mathbf{V}_2 = 0 \Rightarrow k_i \hat{u}_j^2 - k_j \hat{u}_i^2 \Rightarrow \mathbf{k} \times \mathbf{V}_2 = 0$$

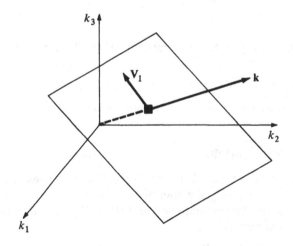

That is, $\mathbf{V}_2$ is collinear with $\mathbf{k}$. For brevity we can say that incompressible turbulent fields belong to the first *mode*. The functional subspace that corresponds to solutions of the Navier–Stokes equations supplemented by $\operatorname{div} \mathbf{V} = 0$ takes on a geometrical meaning in Fourier space. The second mode, which corresponds to compressible effects, is located in the complementary subspace of the Fourier space. It corresponds to a wave mode that is driven both by the random character of the source terms and the random character of the medium in which wave propagation takes place.

From equations

$$\frac{\partial u_i}{\partial t} + u_j \frac{\partial u_i}{\partial x_j} + \frac{1}{\rho} \frac{\partial p}{\partial x_i} = \frac{\lambda + \mu}{\rho} \frac{\partial}{\partial x_i}\left(\frac{\partial u_j}{\partial x_j}\right) + \frac{\mu}{\rho} \frac{\partial^2 u_i}{\partial x_j \partial x_j} \qquad (3.24)$$

$$\underbrace{\qquad\qquad\qquad\qquad\qquad\qquad}_{D(u)}$$

and

$$\frac{\partial \rho}{\partial t} + u_j \frac{\partial \rho}{\partial x_j} + \rho \frac{\partial u_j}{\partial x_j} = 0$$

it is clear that viscous forces play a typical role in each mode. This assumption is supported by experimental data that show the existence of two spectra. The range of the irrotational mode is more extended in wave-number space than the rotational mode, which is usually called *turbulence*. This fact is supported by the momentum equation, put in the following form:

$$D(\mathbf{V}) = \frac{\lambda + \mu}{\rho} \text{grad}[\text{div } \mathbf{V}] + \frac{\mu}{\rho} \Delta \mathbf{V}$$

$$= -\frac{\mu}{\rho} \text{curl}[\text{curl } \mathbf{V}] + \frac{(\lambda + 2\mu)}{\rho} \text{grad}[\text{div } \mathbf{V}]$$

for incompressible   for compressible
mode (turbulence)   mode (waving mode)

We expect the two processes to differ fundamentally, given that the viscosity coefficients are not the same. This difference leads us to consider two modes corresponding to these processes in the physical space. The evolution of each mode is governed by a typical equation. This approach can be more effective than a similar treatment carried out in a spectral space because of the nonlinear mechanisms that introduce convolution products in spectral space. In the relation between p and ρ (e.g., $p = k\rho\gamma$) an exponent term γ appears that introduces many complications, which are all the more difficult to evaluate because the exponent also depends on the dissipative process itself.

This process begins with a transformation of the advective term, which can be written as

$$u_j \frac{\partial u_i}{\partial x_j} = u_j\left(\frac{\partial u_i}{\partial x_j} - \frac{\partial u_j}{\partial x_i}\right) + u_j \frac{\partial u_j}{\partial x_i}$$

$$= -2u_j \varepsilon_{ijl}\omega_l + \frac{1}{l} \frac{\partial}{\partial x_i}(u_j u_j) \qquad (3.25)$$

This splitting is used in a vectorial form to write Bernoulli's equation.

The pressure term yields

$$\frac{1}{\rho}\frac{\partial p}{\partial x_i} = \frac{\partial}{\partial x_i}\left(\frac{p}{\rho}\right) - p\frac{\partial}{\partial x_i}\left(\frac{1}{\rho}\right) \tag{3.26}$$

and the dissipative term displays the role of each mode,

$$\frac{\mu}{\rho}\,\text{curl[curl }\mathbf{V}] - \frac{\lambda + 2\mu}{\rho}\frac{\partial}{\partial x_i}\left[\frac{\partial u_i}{\partial x_j}\right]$$

in which the viscosity coefficients are assumed to be constant.

The curl operator, applied to the momentum equation, leads to

$$\frac{1}{2}\,\text{curl}\left[u_j\frac{\partial u_j}{\partial x_i}\right] \Rightarrow \varepsilon_{ijl}\frac{\partial}{\partial x_j}\left[-2\varepsilon_{lnm}u_n\omega_m\right]$$

$$= -\varepsilon_{ijl}\varepsilon_{lnm}\frac{\partial}{\partial x_j}\left[u_n\omega_m\right]$$

$$= \{\delta_{in}\delta_{jm} - \delta_{im}\delta_{jn}\}\frac{\partial}{\partial x_j}\left[u_n\omega_m\right]$$

$$= -\frac{\partial}{\partial x_j}\left[u_i\omega_j\right] + \frac{\partial}{\partial x_j}\left[u_j\omega_i\right]$$

$$= -\frac{\partial}{\partial x_j}\omega_j + \omega_i\frac{\partial u_j}{\partial u_i} + u_j\frac{\partial\omega_i}{\partial x_j}$$

but it should noted that

$$\frac{1}{2}\,\text{curl}\left[\frac{1}{\rho}\frac{\partial p}{\partial x_i}\right]$$

can be expressed as

$$\frac{1}{2}\varepsilon_{ijl}\frac{\partial}{\partial x_j}\left[p\frac{\partial}{\partial x_l}\left(\frac{1}{\rho}\right)\right] = -\frac{1}{2}\varepsilon_{ijl}\frac{\partial p}{\partial x_j}\frac{\partial}{\partial x_l}\left[\frac{1}{\rho}\right]$$

The viscous term for the irrotational mode is given by

$$\frac{1}{2}\,\text{curl}\left[\frac{2\mu}{\rho}\,\text{curl }\mathbf{\Omega}\right] = -\frac{\mu}{\rho}\Delta\mathbf{\Omega}$$

Finally, the following two equations govern the evolution of $\mathbf{\Omega}$ and C, respectively. The first is not very different from Helmoltz equation except for the choice of the dependent function $\mathbf{\Omega}$ (and not $\mathbf{\Omega}/\rho$).

$$\frac{\partial\omega_i}{\partial t} + u_j\frac{\partial\omega_i}{\partial x_j} + \omega_i C - \frac{\partial u_i}{\partial x_j}\omega_j - \frac{1}{2}\varepsilon_{ijl}\frac{\partial p}{\partial x_j}\frac{\partial}{\partial x_l}\left(\frac{1}{\rho}\right) - \frac{\mu}{\rho}\Delta\omega_i = 0 \tag{3.27}$$

The equation for C can be written

$$\frac{\partial C}{\partial t} + u_j \frac{\partial}{\partial x_j} + \frac{\partial u_j}{\partial x_i} \frac{\partial u_i}{\partial x_j} + \frac{\partial}{\partial x_i}\left(\frac{1}{\rho}\right)\frac{\partial p}{\partial x_i} + \frac{1}{\rho}\Delta p - \frac{\lambda + 2\mu}{\rho}\Delta C = 0$$

with

$$\frac{\partial \rho}{\partial t} + u_j \frac{\partial \rho}{\partial x_j} + \rho C = 0 \qquad \lambda + 2\mu = \frac{4}{3}\mu$$

The two modes are strongly coupled, and this contributes to the randomization of the wavy mode. It we assume with Stokes that $3\lambda + 2\mu = 0$, then the viscous terms are expected to be of different magnitudes.

The equation pertaining to C is obtained by applying the divergence operator to the fundamental equation. At this point, it is worthwhile to introduce the deviatoric tensor in the expression of $D(V)$ such that

$$D(V) = \left[\lambda + \frac{2\mu}{3}\right]\frac{\partial}{\partial x_i}\,\text{div }V + 2\mu\frac{\partial}{\partial x_k}D_{ki}$$

The tensor

$$d_{ij} = D_{ij} - \frac{1}{3}\text{div}V\delta_{ij} \tag{3.28}$$

is the deviatoric tensor (whose divergence is null); applied to the last expression, it leads to

$$D(V) = \left[\lambda + \frac{2\mu}{3}\right]\frac{\partial}{\partial x_i}\,\text{div }V + 2\mu\frac{\partial}{\partial x_k}D_{ki} \tag{3.29}$$

This expression must be compared with that of the dissipation function

$$\Phi = \left[\lambda + \frac{2\mu}{3}\right][\text{div }V]^2 + 2\mu\,d_{ij}\,d_{ij} \tag{3.30}$$

We see that the Stokes assumption is justified both by the necessity to correlate the trace of the stress tensor to the thermodynamic pressure and to the kinetic theory of monoatomic gases. In fact, $D(V)$ and Φ can be rewritten as

$$D(V) = 2\mu\frac{\partial}{\partial x_k}d_{ki} \qquad \text{and} \qquad \Phi = 2\mu\,d_{ij}\,d_{ij} \tag{3.31}$$

Although the compressible part of the field does not contribute to the dissipation that accounts for the Stokes assumption $3\lambda + 2\mu = 0$; in the rate equation for the divergence operator, some dissipation terms reappear. These can be expressed as

$$\text{div } \mathbf{D}(\mathbf{V}) = \text{div}\left(2\mu \frac{\partial}{\partial x_k} d_{ki}\right)$$

$$= 2\mu \frac{\partial}{\partial x_i} \frac{\partial}{\partial x_k} D_{ki} - \frac{2}{3} \mu \frac{\partial}{\partial x_i} \frac{\partial}{\partial x_k} \text{div } \mathbf{V}\delta_{ki}$$

$$= 2\mu \frac{\partial}{\partial x_i} \frac{\partial}{\partial x_k} \frac{1}{2}\left[\frac{\partial u_i}{\partial x_k} + \frac{\partial u_k}{\partial x_i}\right] - \frac{2}{3} \mu \frac{\partial^2}{\partial x_i \partial x_j} \text{div } \mathbf{V}$$

$$= \mu \frac{\partial}{\partial x_k \partial x_k} \frac{\partial u_i}{\partial x_i} + \left[\frac{\partial^2}{\partial x_i \partial x_i} \frac{\partial u_k}{\partial x_k}\right] - \frac{2}{3} \mu \frac{\partial^2}{\partial x_i \partial x_i} \text{div } \mathbf{V}$$

$$= \left[2\mu - \frac{2}{3} \mu\right] \Delta(\text{div } \mathbf{V})$$

$$\text{div } \mathbf{D}(\mathbf{V}) = \frac{4}{3} \mu\Delta(\text{div } \mathbf{V})$$

This result is worthy of comment. With Stokes's assumption, the part of the dissipation related to dilatation effects becomes null. At first look, we could expect that viscous terms are eliminated in all the equations dealing with compressible effects. Previous computation, however, shows this conclusion to be wrong. The divergence operator, applied to the deviatoric tensor **D** whose trace is null, reintroduces the divergence of **V**, which is supposed to be different from zero. This clearly shows how important the level of description of a phenomenon concerning the role of typical terms really is.

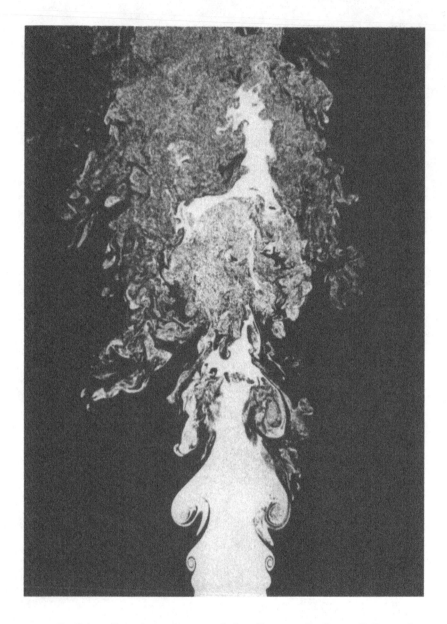

Radial section of an axisymmetric jet. Courtesy G. Comte-Bellot and Laboratoire de Mécanique des Fluides, Ecole Centrale de Lyon, Ecully, France.

4 Fundamental Concepts of Turbulence

This chapter provides basic information on the turbulence field to give the reader an understanding of the usual laws. For instance, *the law of the wall* supposes that the turbulence near the wall is more or less independent of that in the outer part of the boundary layer. But is it? Where is the turbulence produced? In what manner is the turbulent energy distributed throughout the flow? What is the effect of the turbulence on the mean velocity field? Even though we are not chiefly interested in the behavior of turbulent fields, these and similar questions must be answered.

First, we examine these fundamental questions: What are the origins of the turbulence and what external parameters dominate its generation? Extensive work has been carried out on this subject, which is more or less connected with problems of stability. In a first step, the Navier–Stokes equations are linearized. The development of a small disturbance applied to a basic velocity field is then examined. Does this disturbance decrease or increase with time? If it decreases, the basic flow should be considered stable; if it increases, the basic flow becomes unstable, and the laminar initial flow could become turbulent. In the *method of small oscillations*, only disturbances that are compatible with the equations of motion are retained. The manner in which these disturbances develop in the flow is analyzed with the aid of the equations of motion. Here we apply the method of small disturbances to the case of a two-dimensional mean flow.

We assume the basic velocity field to be dependent only on the ordinate y, whereas the V^B component of the velocity is taken to be null everywhere. Flows that fulfill these conditions are termed *parallel flows*. They represent a good approximation of boundary layer flows because the dependence of U^B on the x coordinate is much smaller than that on y. In the case of a channel with parallel walls, a parallel flow is produced with great accuracy at a sufficient distance from the inlet station.

The velocity and pressure describing the basic field are

$$U^B(y) \qquad V^B \approx 0 \qquad P^B(x, y)$$

If we describe a two-dimensional disturbance as

$$\tilde{u}(x, y, t) \qquad \tilde{v}(x, y, t) \qquad \tilde{p}(x, y, t)$$

the global field can be inferred to be

$$U = U^B + \tilde{u} \qquad V = \tilde{v} \qquad P = P^B + \tilde{p}$$

From the linearity assumption, we can see that the two fields may be considered as decoupled. Each of them is supposed to verify the Navier–Stokes equations, so we write

$$\frac{\partial \tilde{u}}{\partial t} + U \frac{\partial \tilde{u}}{\partial x} + \tilde{u} \frac{\partial \tilde{u}}{\partial x} + \tilde{v} \frac{DU}{dy} + \tilde{v} \frac{\partial \tilde{u}}{\partial y}$$

$$= -\frac{1}{\rho} \frac{\partial P}{\partial x} - \frac{1}{\rho} \frac{\partial \tilde{p}}{\partial x} + \nu \left[\frac{d^2 U}{dy^2} + \frac{\partial^2 \tilde{u}}{\partial x^2} + \frac{\partial^2 \tilde{u}}{\partial y^2} \right] \qquad (4.1)$$

$$\frac{\partial \tilde{v}}{\partial t} + U \frac{\partial \tilde{v}}{\partial x} + \tilde{v} \frac{\partial \tilde{v}}{\partial y} = -\frac{1}{\rho} \frac{\partial P}{\partial y} - \frac{1}{\rho} \frac{\partial \tilde{p}}{\partial y} + \nu \left[\frac{\partial^2 \tilde{v}}{\partial x^2} + \frac{\partial^2 \tilde{v}}{\partial y^2} \right] \qquad (4.2)$$

$$\frac{\partial \tilde{u}}{\partial x} + \frac{\partial \tilde{v}}{\partial y} = 0 \qquad (4.3)$$

With the assumptions made for the basic field, the continuity equation is obviously satisfied. And because the main field verifies the Navier–Stokes equations, we can write, after subtraction,

$$\begin{cases} \dfrac{\partial \tilde{u}}{\partial t} + U \dfrac{\partial \tilde{u}}{\partial x} + \tilde{v} \dfrac{\partial U}{\partial y} = -\dfrac{1}{\rho} \dfrac{\partial \tilde{p}}{\partial s} + \nu \left[\dfrac{\partial^2 \tilde{u}}{\partial x^2} + \dfrac{\partial^2 \tilde{u}}{\partial y^2} \right] \\[3mm] \dfrac{\partial \tilde{v}}{\partial y} + U \dfrac{\partial \tilde{v}}{\partial x} + \tilde{v} \dfrac{\partial \tilde{v}}{\partial y} = -\dfrac{1}{\rho} \dfrac{\partial \tilde{p}}{\partial y} + \nu \left[\dfrac{\partial^2 \tilde{v}}{\partial x^2} + \dfrac{\partial^2 \tilde{v}}{\partial y^2} \right] \end{cases}$$

$$\frac{\partial \tilde{u}}{\partial x} + \frac{\partial \tilde{v}}{\partial y} = 0$$

These equations must obviously be supplemented by the initial values for the disturbances and the boundary conditions, which require that u and v be zero at the wall (no-slip condition). Assuming the basic field to be a given function of y that satisfies the no-slip condition at the wall, we finally have the three unknown functions

$$\tilde{u} \qquad \tilde{v} \qquad \tilde{p}$$

to solve for with the previous three equations.

It is easy to eliminate the pressure $\tilde{p}$ in the first two equations and to obtain the two unknown functions $\tilde{u}$ and $\tilde{v}$ from the two remaining equations. The perturbation being two-dimensional, we can define a stream function

$$\Psi(x, y, t) \quad \begin{cases} \tilde{u} = \dfrac{\partial \Psi}{\partial y} \\[2ex] \tilde{v} = -\dfrac{\partial \Psi}{\partial x} \end{cases}$$

that satisfies the continuity equation.

The disturbance stream function is assumed to be of the form

$$\Psi(x, y, t) = \Phi(y) \, e^{i(\alpha x - \beta t)} \tag{4.4}$$

in which α is a real quantity, and the wavelength λ is given by $\lambda = 2\pi/\alpha$. As for β, it is the complex quantity $\beta = \beta_r + i\beta_i$ defining the circular frequency of the partial oscillation in which the imaginary part determines the degree of amplification or damping. When it is negative, the disturbances are damped; otherwise an instability sets in. The stream function can be written as

$$\begin{aligned} \Psi(x, y, t) &= \Phi(y) \, e^{i\alpha[x - (\beta/\alpha)t]} \\ &= \Phi(y) \, e^{i\alpha[x - (c_r + ic_i)t]} \end{aligned}$$

with

$$c = \frac{\beta}{\alpha} = c_r + ic_i$$

in which c_r denotes the propagation of the wave in the x direction (phase velocity). Finally, the equation for $\Phi(y)$ can be written, if we assume $\nabla^2(u + U_B) \sim \nabla^2 U_B$ as

$$[U^B - c][\Phi'' - \alpha^2 \Phi] - (U^B)'' \Phi = -\frac{i\nu}{\alpha}[\Phi'' - 2\alpha^2 \Phi'' + \alpha^4 \Phi] \tag{4.5}$$

It is convenient now to introduce dimensionless variables by dividing all velocities by the maximum velocity U_m of the basic field and by dividing all lengths by a suitable selected length such as the boundary layer thickness. We denote differentiations with respect to the dimensionless quantity y/δ by primes. The Reynolds number $Re = U_m\delta/\nu$ appears in the final equation

$$[U_B - c][\Phi'' - \alpha^2 \Phi] - U_B'' \Phi = -\frac{i}{\alpha Re}[\Phi'''' - 2\alpha^2 \Phi'' + \alpha^4 \Phi] \tag{4.6}$$

which is supplemented by the following boundary conditions:

$$\begin{cases} y = 0 & \tilde{u} = \tilde{v} = 0 \qquad \Phi = 0 \qquad \Phi' = 0 \\ y = \infty & \tilde{u} = \tilde{v} = 0 \qquad \Phi = 0 \qquad \Phi' = 0 \end{cases}$$

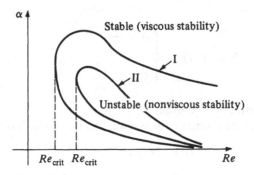

Fig. 4.1 Stability curves

This fourth-order differential equation, together with the boundary conditions, leads to an eigenvalue problem. Nontrivial values of Φ can be found when c has selected values only, so the problem becomes completely defined. A solution exists for selected values of the propagation velocity of the wave only.

We shall not detail the method for carrying out the computations, but we shall present the main results and make some comments on the shortcomings of the method when it is considered as a prediction method for the transition from laminar to turbulent flows. All the results are shown in the Fig. 4.1, where the coordinates are the Reynolds number $Re = U_m^B \delta / \nu$ and the wavelength of the disturbance (α or λ).

For a given Reynolds number, disturbances exhibiting typical wave numbers are unstable; that is, they grow with time. An area that corresponds to unstable disturbances is defined in the plane (α, Re). The limits of this area depend on the terms taken into account: For viscous stability we have curve I, and for nonviscous stability we have curve II.

Remarks: The previous computation can be extended to more complex two-dimensional waves by using Fourier series and Fourier integrals. In this computation, both the disturbance and the basic flow are assumed to be two-dimensional. These remarks are in perfect agreement with each other; however, they greatly reduce the ability of the method to predict the transition from laminar to turbulent flow. Such computations provide some directions but no precise information.

The method introduces preselected functions. Because the form of the equation under consideration becomes linear, we expect an exponential solution. Computations have been made that do not presume the shape of the function and give results similar to those obtained by using specified functions.

Unfortunately, the methods for treating the problem of linear stability cannot be extended to the prediction of turbulence because the transition starts with turbulent spots that increase in size as they are convected downstream. The phenomenon is not continuous. Many hypotheses can be made regarding the problem of turbulence emanating from nonlinear equations. How such a system can evolve from deterministic behavior to random and intermittent behavior has always been an open question, and descriptions of paths to chaos have been developed only recently.

4.2 A STATISTICAL APPROACH TO TURBULENT FLOW

4.2.a *Many ways to handle the problem*

A statistical way to obtain useful information on the role of turbulence on mean velocity profiles was introduced by Reynolds at the end of the last century. The turbulent flow is separated into two parts, a mean flow characterized by $\overline{U_i}$, $\overline{P}$ and a turbulent part characterized by u_i, p. The quantities describing the whole flow have been termed U_i and P, which can be decomposed as follows:

$$\begin{cases} \overline{U_i}, \overline{P} & U_i = \overline{U_i} + u_i \\ u_i, p & P = \overline{P} + p \end{cases}$$

Here we assume the fluid to be incompressible so that the corresponding splitting is not necessary for the density ρ and the temperature T. The Navier–Stokes equations are assumed to be verified for the whole flow.

All the information contained in the starting equation can be restored only by an infinite set of equations relative to the statistical moments that are directly introduced by this method. Even though the problem is completely determined by the Navier–Stokes equations, an open problem appears at each step when a statistical method is applied to treat a nonlinear problem.* The global information can be provided only through an infinite set of coupled equations. If such detailed information is needed, it is better to deal with the Navier–Stokes equations themselves, assuming that they have a unique solution when associated with convenient initial and boundary conditions. For technical problems it is often possible to be content with a reduced set of information; however, it is difficult to predict the consequences of truncating the problem.

---

* For a linear problem, equations governing statistical moments are not coupled. Information can be searched for separately, but obtaining all the information also requires solving an infinite set of uncoupled equations.

4.2.b *Other ways*

To obtain all the information, full numerical simulations can be used. Because the computing time required is long, such simulations can only be applied to very simple cases (i.e., homogeneous fields).

4.2.c *The mean velocity field*

To use a statistical method we must write the equation governing the mean velocity field. With the usual splitting, the continuity equation can be written

$$\frac{\partial U_i}{\partial x_i} = \frac{\partial [\overline{U}_i + u_i]}{\partial x_i} = 0$$

which (after averaging and taking the difference) leads to

$$\frac{\partial \overline{U}_i}{\partial x_i} = 0 \qquad \frac{\partial u_i}{\partial x_i} = 0$$

The Navier–Stokes equations are then

$$\frac{\partial [\overline{U}_i + u_i]}{\partial t} + (\overline{U}_j + u_j) \frac{\partial [\overline{U}_i + u_i]}{\partial x_j}$$

$$= -\frac{1}{\rho} \frac{\partial [\overline{P} + p]}{\partial x_i} + \nu \frac{\partial^2 [\overline{U}_i + u_i]}{\partial x_j\, \partial x_j} \qquad (4.7)$$

which after averaging become

$$\frac{\partial \overline{U}_i}{\partial t} + \overline{U}_j \frac{\partial \overline{U}_i}{\partial x_j} + \overline{u_j \frac{\partial u_i}{\partial x_j}} = -\frac{1}{\rho} \frac{\partial \overline{P}}{\partial x_i} + \nu \frac{\partial^2 \overline{U}_i}{\partial x_j\, \partial x_j}$$

advective terms due to
mean and turbulent flows

This equation can be rewritten as

$$\frac{\partial \overline{U}_i}{\partial t} + \overline{U}_j \frac{\partial \overline{U}_i}{\partial x_j} = \frac{\partial}{\partial x_j} \left\{ -\overline{u_i u_j} - \frac{\overline{P}}{\rho} \delta_{ij} + \nu \left(\frac{\partial \overline{U}_i}{\partial x_j} + \frac{\partial \overline{U}_j}{\partial x_i} \right) \right\}$$

This form refers to advective terms due to the mean velocity field only, whereas those due to the turbulent motion are interpreted as an additional stress. The straining process linked to the mean motion is also emphasized:

$$\frac{\partial \overline{U}_i}{\partial t} + \overline{U}_i \frac{\partial \overline{U}_i}{\partial x_i} = \frac{1}{\rho} \frac{\partial}{\partial x_j} \left\{ \underbrace{-\rho \overline{u_i u_j}}_{\text{Reynolds stress tensor}} - \overline{P} \delta_{ij} + \underbrace{\mu \left(\frac{\partial \overline{U}_i}{\partial x_j} + \frac{\partial \overline{U}_j}{\partial x_i} \right)}_{\text{viscous stress tensor}} \right\} \qquad (4.8)$$

The nonlinearity of the initial problem introduced a coupling process between the mean velocity field and the Reynolds stress tensor (second-order correlations at one point). In the region of the flow where the turbulence motion is fully developed, $\rho \overline{u_i u_j}$ can be as much as two or three orders of magnitude greater than $\mu \, \partial \overline{U}_i / \partial x_j$.

$$-\rho \overline{u_i u_j} \gg \mu \left(\frac{\partial \overline{U}_i}{\partial x_j} + \frac{\partial \overline{U}_j}{\partial x_i} \right)$$

To close the problem, the introduction of a first gradient approximation is the simplest assumption we can make. This is expressed by

$$-\overline{u_i u_j} = \nu_T \left(\frac{\partial \overline{U}_i}{\partial x_j} + \frac{\partial \overline{U}_j}{\partial x_i} \right) \tag{4.9}$$

in which ν_T is an eddy coefficient somewhat similar to the kinematic viscosity coefficient. This assumption, however, appears to be inconsistent when we examine some simple cases.

$$\left. \begin{aligned} i = j = 1 &\Rightarrow -\overline{u_1^2} = 2\nu_T \frac{\partial \overline{U}_1}{\partial x_1} \\[2mm] i = j = 2 \quad &\Rightarrow -\overline{u_2^2} = 2\nu_T \frac{\partial \overline{U}_2}{\partial x_2} \\[2mm] i = j = 3 &\Rightarrow -\overline{u_3^2} = 2\nu_T \frac{\partial \overline{U}_3}{\partial x_3} \end{aligned} \right\} \Rightarrow \overline{u_1^2} + \overline{u_2^2} + \overline{u_3^2} = 0$$

We can make a slight improvement by introducing the trace of the Reynolds stress tensor. In this case,

$$\sum_{i=j} \overline{u_i u_j} = \overline{u_1^2} + \overline{u_2^2} + \overline{u_3^2} = \overline{q^2}$$

$$\left. \begin{aligned} -\overline{u_1^2} + \frac{1}{3} \overline{q^2} &= 2\nu_T \frac{\partial \overline{U}_1}{\partial x_1} \\[2mm] -\overline{u_2^2} + \frac{1}{3} \overline{q^2} &= 2\nu_T \frac{\partial \overline{U}_2}{\partial x_2} \\[2mm] -\overline{u_3^2} + \frac{1}{3} \overline{q^2} &= 2\nu_T \frac{\partial \overline{U}_3}{\partial x_3} \end{aligned} \right\} \Rightarrow \overline{u_1^2} = \overline{u_2^2} = \overline{u_3^2}$$

Although the turbulence appears to be isotropic in the case of a channel flow (with turbulent kinetic energy $\overline{q^2}$), in most industrial cases the turbulence exhibits a high degree of anisotropy. These inconsistencies are the consequence of two features: (a) Turbulent transfers are not local and (b) the principal axes of the Reynolds stress tensor do not generally coincide with the principal axes of the straining tensor for shear flows. In fact, the use of statistical methods introduces new quantities (correlation terms $\overline{u_i u_j}$), and we have to deal with them instead of the instantaneous functions describing the state of the field at each time and location (U_i, P).

After examining the behavior of the correlation terms, we expect such quantities to introduce explicitly a kind of memory: The correlation terms are not instantaneously responsive to the mean velocity gradients. In this form the problem becomes nonlocal and requires some ad hoc treatment. In using a statistical approach, the basic problem is to know how to introduce consistent assumptions. The degree of sophistication required for making such assumptions is to be expected in view of the close connection that exists with the order of the statistical moment. This can be conjectured, but it is not evident at all. If we accept this point of view, we want to go farther and write the equation governing the double velocity correlations $\overline{u_i u_j}$.

4.2.d *Equations for the Reynolds stress tensor and the turbulent kinetic energy*

We start again by applying the Navier–Stokes equations to the instantaneous flow:

$$\frac{\partial}{\partial t}(\overline{U_i} + u_i) + (\overline{U_k} + u_k)\frac{\partial}{\partial x_k}[\overline{U_i} + u_i] = -\frac{1}{\rho}\frac{\partial(\overline{P} + p)}{\partial x_i} + \nu\frac{\partial^2(\overline{U_i} + u_i)}{\partial x_k \partial x_k}$$

which, after averaging and taking into account the continuity equation, become

$$\frac{\partial \overline{U_i}}{\partial t} + \overline{U_k}\frac{\partial \overline{U_i}}{\partial x_k} + \frac{\partial}{\partial x_k}\overline{u_i u_k} = -\frac{1}{\rho}\frac{\partial \overline{P}}{\partial x_i} + \nu\frac{\partial^2 \overline{U_i}}{\partial x_k \partial x_k}$$

Subtracting this equation from the first one, we obtain

$$\frac{\partial u_i}{\partial t} + \overline{U_k}\frac{\partial u_i}{\partial x_k} + u_k\frac{\partial \overline{U_i}}{\partial x_k} + \frac{\partial}{\partial x_k}[u_i u_k - \overline{u_i u_k}] = -\frac{1}{\rho}\frac{\partial p}{\partial x_i} + \nu\frac{\partial^2 u_i}{\partial x_k \partial x_k}$$

and for u_j a similar equation can be inferred:

$$\frac{\partial u_j}{\partial t} + \overline{U_k}\frac{\partial u_j}{\partial x_k} + u_k\frac{\partial \overline{U_j}}{\partial x_k} + \frac{\partial}{\partial x_k}[u_j u_k - \overline{u_j u_k}]$$

$$= -\frac{1}{\rho}\frac{\partial p}{\partial x_j} + \nu\frac{\partial^2 u_j}{\partial x_k \partial x_k} \qquad (4.10)$$

After these last two equations are multiplied by u_j and u_i, respectively, added, and averaged, we find

$$2\frac{\partial \overline{u_i u_j}}{\partial t} + \overline{U_k}\frac{\partial}{\partial x_k}\,\overline{u_i u_j} = -\left(\overline{u_j u_k}\frac{\partial \overline{U_i}}{\partial x_k} + \overline{u_i u_k}\frac{\partial \overline{U_j}}{\partial x_k}\right) - \frac{\partial}{\partial x_k}\,\overline{u_i u_j u_k}$$

$$-\frac{1}{\rho}\left\{\overline{u_j\frac{\partial p}{\partial x_i}} + \overline{u_i\frac{\partial p}{\partial x_j}}\right\}$$

$$+\nu\left\{\overline{u_j\frac{\partial^2 u_i}{\partial x_k\,\partial x_k}} + \overline{u_i\frac{\partial^2 u_j}{\partial x_k\,\partial x_k}}\right\}$$

The two terms on the left-hand side represent unsteady and advective effects, respectively, whereas the terms on the right-hand side represent, respectively, the interaction between the mean and the turbulent flow, the advective effects of the turbulent motion, the pressure effects related to the turbulent kinematic field, and the viscous effects that generate molecular diffusion and dissipation. After setting $i = j$ and summing with respect to the repeated index, we obtain

$$2\frac{\partial}{\partial t}\,\overline{q^2} \Rightarrow \text{unsteady effects}$$

$$+\,\overline{U_k}\frac{\partial \overline{q^2}}{\partial x_k} \Rightarrow \text{advection by the mean motion}$$

$$= -2\overline{u_i u_k}\frac{\partial \overline{U_i}}{\partial x_k} \Rightarrow \begin{cases}\text{interaction between the turbulent and}\\ \text{the mean field (production term)}\end{cases}$$

$$-\frac{\partial}{\partial x_k}\,\overline{q^2 u_k} \Rightarrow \text{advection by the fluctuation motion}$$

$$-2\frac{1}{\rho}\frac{\partial}{\partial x_i}\,\overline{p\overline{u_i}} \Rightarrow \text{pressure effects}$$

$$+2\nu\overline{u_i\frac{\partial^2 u_i}{\partial x_k\,\partial x_k}} \Rightarrow \text{viscous effect} \begin{cases}\text{molecular diffusion}\\ \text{dissipation}\end{cases}$$

For the homogeneous perturbation field, the space derivatives of mean quantities are obviously null, so the turbulent kinetic equation can be written as

$$\frac{\partial}{\partial t}\,\overline{q^2} = -\overline{u_i u_k}\frac{\partial \overline{U_i}}{\partial x_k} - (\text{dissipation term})$$

      evolution   production   dissipation

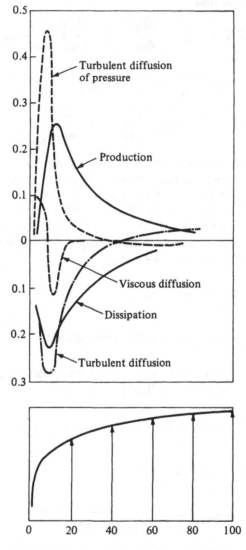

Fig. 4.2   Turbulent kinetic energy balance

At this stage, the most important feature is the location of the production regions. For example, in a boundary layer a unique production
center is located next to the wall (Fig. 4.2). On average, information
goes from the inner region to the outer region of the boundary layer,
so we can infer that the wall region depends on local parameters only.
Note that the pressure terms disappear when passing from the equation

for $\overline{u_i u_j}$ to that for $\overline{q^2}$. At this point, we can anticipate the role of the pressure term in the directional distribution of the turbulent kinetic energy. (The contributions of $\overline{u_1^2}$, $\overline{u_2^2}$, and $\overline{u_3^2}$ to $\overline{q^2}$ are not equal except for an isotropic turbulence.) The production term is very large near the wall; for a two-dimensional flow it corresponds to

$$\overline{u_1 u_2} \frac{d\overline{U_1}}{dx_2}$$

where both the stress term $\overline{u_1 u_2}$ and the mean velocity gradient are large. The turbulent bulges are rather small, so the dissipation is very important. Because a quasi-equilibrium exists, local interactions are prevalent.

4.2.e   *Isotropic turbulence, subjected to a shear flow*

Now we extend this kind of explanation to an isotropic turbulence subjected to a shear flow. We use again the assumption of homogeneity so that all the space derivatives of statistical quantities linked to the turbulent field are null. Whenever a field is assumed to be homogeneous, it is of course unbounded. From this point of view our approach is unrealistic, but it is nevertheless useful for understanding the mechanisms that come into play in more complex flows such as boundary layers, jets, wall jets, and mixing zones.

The mean field in the case under consideration, is not altered by the turbulent motion; in fact, all the space derivatives of the Reynolds stress tensor are null:

$$\frac{\partial}{\partial x_k} \overline{u_i u_k} = 0 \tag{4.11}$$

The turbulent field does not appear in the equation dealing with the mean velocity field. Thus, the mean velocity gradient, which acts on the turbulent motion, is not altered by this random motion, and so the pattern of interaction between the two fields can be viewed as a one-way street. The problem is greatly simplified but not distorted. Mechanisms that orient the turbulent structures are at play here, and we can see them typically acting in the equation for $\overline{u_i u_j}$, which can be written in its simplified form:

$$2\frac{\partial \overline{u_i u_j}}{\partial t} = -\left\{\overline{u_j u_k}\frac{\partial \overline{U_i}}{\partial x_k} + \overline{u_i u_k}\frac{\partial \overline{U_j}}{\partial x_k}\right\} - \frac{1}{\rho}\left\{\overline{u_j \frac{\partial p}{\partial x_i}} + \overline{u_i \frac{\partial p}{\partial x_j}}\right\} + \text{(viscous terms)}$$

This equation is given in this equivalent form:

$$2\frac{\partial \overline{u_i u_j}}{\partial t} = -\left\{\overline{u_j u_k}\frac{\partial \overline{U_j}}{\partial x_k} + \overline{u_i u_k}\frac{\partial \overline{U_i}}{\partial x_k}\right\} + \frac{1}{\rho}\left\{\overline{p\left(\frac{\partial u_j}{\partial x_i} + \frac{\partial u_i}{\partial x_j}\right)}\right\} + \text{(viscous terms)}$$

Because the field is assumed to be homogeneous, we can write

$$\frac{\partial}{\partial x_i}\,\overline{pu_j} = 0 = \overline{p\frac{\partial u_j}{\partial x_i}} + \overline{u_j\frac{\partial p}{\partial x_i}} \Rightarrow \overline{u_j\frac{\partial p}{\partial x_i}} = -\overline{p\frac{\partial u_j}{\partial x_i}}$$

This last form shows the role of the rate-of-strain–pressure correlations associated with the turbulent motion.

If the unique component of the tensor $\partial \overline{U}_i/\partial x_k$ is $\partial \overline{U}_1/\partial x_2$, then

$$2\frac{\partial \overline{u_1 u_2}}{\partial t} = -\overline{u_2^2}\frac{\partial \overline{U}_1}{\partial x_2} + \frac{1}{\rho}\overline{\left\{p\left(\frac{\partial u_1}{\partial x_2} + \frac{\partial u_2}{\partial x_1}\right)\right\}} \qquad (4.12)$$

for this component viscous effects are negligible

The evolution of this component of the Reynolds stress tensor is governed by two terms. The first is closely related to the mean velocity gradient $\partial \overline{U}_1/\partial x_2$; the second shows that the role of the pressure term is far more complex. To get more details, we rewrite the rate equation for the instantaneous motion in the form

$$\frac{\partial}{\partial t}(\overline{U}_i + u_i) + \frac{\partial}{\partial x_j}[(\overline{U}_j + u_j)(\overline{U}_i + u_i)] = -\frac{1}{\rho}\frac{\partial(\overline{P} + p)}{\partial x_i} + \nu\frac{\partial^2(\overline{U}_i + u_i)}{\partial x_j \partial x_j}$$

and to elucidate the role of the pressure term we take the divergence of the rate equation:

$$-\frac{1}{\rho}\Delta(\overline{P} + p) = \frac{\partial^2}{\partial x_i \partial x_j}(\overline{U}_i + u_i)(\overline{U}_j + u_j) \qquad (4.13)$$

The mean and the fluctuating parts of the pressure are connected with the whole velocity field through the two equations (obtained as described later), which after averaging become

$$-\frac{\Delta \overline{P}}{\rho} = \frac{\partial^2}{\partial x_i \partial x_j}\{\overline{U}_i \overline{U}_j + \overline{u_i u_j}\}$$

and, subtracting from the initial equation, we obtain

$$\frac{\Delta p}{\rho} = -\frac{\partial^2}{\partial x_i \partial x_j}\{\overline{U}_i u_j + u_i \overline{U}_j + u_i u_j - \overline{u_i u_j}\}$$

After some rearrangements and taking into account the continuity equation, the rate equation for the fluctuating pressure can be written

$$-\frac{\Delta p}{\rho} = \underbrace{2\frac{\partial \overline{U}_i}{\partial x_j}\frac{\partial u_j}{\partial x_i}}_{\text{linear term}} + \underbrace{\frac{\partial u_i}{\partial x_j}\frac{\partial u_j}{\partial x_i}}_{\text{non linear term}} \qquad (4.14)$$

Equation (4.14) clearly shows the linear and nonlinear effects. The rate-of-strain–pressure correlation has a global role in restraining the

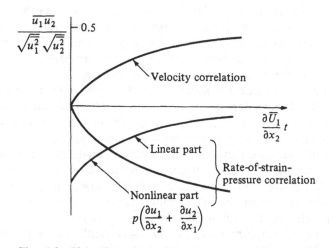

Fig. 4.3   Velocity and rate-of-strain–pressure correlations, showing that an equilibrium state is reached for a moderate value of $(\partial\overline{U}_1/\partial x_2)t$

trend to anisotropy; accordingly, the anisotropy of the turbulent field, which can be measured through the term $\overline{u_1 u_2}$, is the result of a delicate balance between two terms (Fig. 4.3). The first terms $\overline{u_2^2}(\partial\overline{U}_1/\partial x_2)$ increases the anisotropy of the field through the mean velocity gradient, and the second term, which is a pressure term, restrains this trend; so a sort of equilibrium takes place after some delay. The relevant parameter representing this evolution is $(\partial\overline{U}_1/\partial x_2)t$: The turbulent field is strained by the mean field in a cumulative way. In this kind of mechanism, large turbulent structures are involved and not small structures, which are only responsible for dissipation effects. These small structures were assumed in the previous computations to be in an isotropic state so the role of viscosity was removed from the equation to $\overline{u_1 u_2}$.

The linear part of the rate-of-strain–pressure correlations acts as soon as the initial isotropic field is strained (sudden part of the pressure term), whereas the nonlinear part of the pressure term intervenes as the anisotropy of the field develops (Rotta's term).

$$2\frac{\partial\overline{u_1 u_2}}{\partial t} = \underbrace{\overline{u_2^2}\frac{\partial\overline{U}_1}{\partial x_2}}_{\substack{\text{straining process}\\\text{applied by}\\\text{the mean field}}} + \underbrace{\frac{1}{\rho}\overline{p\left[\frac{\partial u_1}{\partial x_2} + \frac{\partial u_2}{\partial u_1}\right]}}_{\substack{\text{linear effects}\\\text{+ nonlinear effects}}} \tag{4.15}$$

$$\underbrace{\phantom{2\frac{\partial\overline{u_1 u_2}}{\partial t} = \overline{u_2^2}\frac{\partial\overline{U}_1}{\partial x_2} + \frac{1}{\rho}\overline{p\left[\frac{\partial u_1}{\partial x_2} + \frac{\partial u_2}{\partial u_1}\right]}}}_{\substack{\text{the role of the mean velocity gradient}\\\text{is restricted by the pressure field}}}$$

The essential features are clearly emphasized in this equation, but we should mention that the details of these mechanisms are far easier to understand by using Fourier transforms. In particular, the linear effects of the pressure terms do not introduce an open problem, so a closure is easily obtained by working in this spectral space.

Most flows of practical importance involve mean velocity gradients, the intensity of which depends on their particular location and the extent of the field, which is limited. The interactions between mean and turbulent fields (a two-way street) render the problem more complex. However, the essential features of mean velocity gradients can be investigated by treating them as homogeneous fields. Many researchers have extensively investigated this particular route by using Fourier transforms.

The method used to obtain the equation for $\overline{u_i u_j}$ corresponds to the following scheme. By calling the Navier–Stokes equation for the instantaneous motion E and that for the mean field $\overline{E}$, an equation for the fluctuating motion is obtained by subtracting $\overline{E}$ from E,

$$e = E - \overline{E}$$

This represents an equation for u_i or u_j. The equation for u_i is termed e_i and that for u_j is e_j. Multiplying e_i by u_j and e_j by u_i and adding, we obtain the equation for $u_i u_j$, which after averaging gives us $\overline{u_i u_j}$.

$$E - \overline{E} \Rightarrow \begin{Bmatrix} e_i & \Rightarrow & e_i u_j \\ e_j & \Rightarrow & e_j u_i \end{Bmatrix} + \Rightarrow \left\{ \begin{matrix} \text{equation} \\ \text{for } u_i u_j \end{matrix} \right. \left\{ \begin{matrix} \text{by averaging} \\ \text{we get } \overline{u_i u_j} \end{matrix} \right.$$

4.3   THE SCALING OF TURBULENCE

4.3.a   *Kinematic properties*

In the previous section, we presented the equations governing the evolution of the main statistical properties of turbulence. These equations are given at a single point in physical space, so no reference can be made to a scaling process. Moreover, some terms cannot be evaluated, and they require adequate assumptions for solving the problem. The linear part of the velocity–pressure correlations requires a closure hypothesis as a consequence of the character of the pressure terms, which play a nonlocal role. The scaling of turbulence is so important in many industrial problems that we feel it is useful to present it succinctly here.

In a boundary layer, the turbulent structures located next to the wall do not have the same characteristic length as those encountered in the outer part of the boundary layer. Whenever chemical reactions occur in a turbulent medium, we can conjecture that the turbulent structures play a role according to their sizes, but in the end, molecular effects control the reaction. If we want to examine some industrial applications, it is necessary to introduce the concept of length scales in a more or less sophisticated way.

We introduce turbulence scaling concepts by starting with the correlations at two points x and x + r:

$$\overline{u_i(\mathbf{x}, t)u_j(\mathbf{x} + \mathbf{r}, t)} = R_{ij}(\mathbf{x}, \mathbf{r}, t)$$

Computations can be cumbersome if R depends on the three variables x, r, and t. The scaling concept can be simplified if homogeneous fields are considered. In that case, any tensor related to the statistical properties of turbulence depends on r and t only and can be expressed as R(r, t).

The properties of the equations that describe homogeneous turbulence fields have been explored thoroughly. If we introduced Fourier transforms, the spectral tensor $\Phi(\mathbf{k}, t)$ is thereby associated with R(r, t). Fourier transforms in vectorial space must be considered as extensions of

$$\phi(n) = \int f(t) \, e^{-int} \, dt$$

By integrating we lose some information that is recovered through a kernel function $e^{-i\mathbf{k}\cdot\mathbf{r}}$ with a vectorial parameter k. By means of $\Phi(\mathbf{k}, t)$ the contribution of turbulent structures to R is analyzed, taking into account the size of these structures and their orientations through a wave number vector. Large structures are associated with small values of $|\mathbf{k}|$, and small structures are associated with high values of $|\mathbf{k}|$.

$$\Phi_{ij}(\mathbf{k}, t) = \frac{1}{(2\pi)^3} \iiint R_{ij}(\mathbf{r}, t) \, e^{-i\mathbf{k}\cdot\mathbf{r}} \, d\mathbf{r}$$

At any given time we can consider the degree of dependence of the velocity components u_i and $u'_j = u_j(\mathbf{x} + \mathbf{r}, t)$ as a function of their separation distance r.

In fact, the tensorial quantity R_{ij} is a function of r and t. We can expect the degree of dependence between u_i and u'_j to decrease monotonically as the distance between the two points increases (see Fig. 4.4). If we disregard some special cases, we find that this behavior is in fairly good agreement with experimental data. This loss of coherence seems to be inherent in the nature of turbulence rather than in the random character of the field. In fact, randomness can be introduced by the random motion of a frontier that limits the field, in which case the induced motion must be considered completely deterministic and the correlation coefficients are equal to 1 at any time.

A turbulence field is characterized by this loss of coherence, which can be measured with respect to space or time variables or with respect

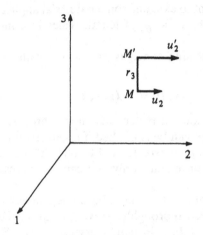

Fig. 4.4   Definition sketch for velocity and separation

to any suitable arrangement made with these same variables (space–time correlations). The correlation coefficient, a dimensionless form of the correlation term $\overline{u_i u_j'}$, is defined as

$$\frac{\overline{u_i u_j'}}{\sqrt{\overline{u_i^2}}\ \sqrt{\overline{u_j'^2}}}$$

The correlation coefficients for the correlation $\overline{u_i u_j'}$ go from one to zero (see Fig. 4.5) as the separation between M and M' increases. A first scaling can be introduced from R, which becomes the Reynolds stress tensors as the two points M and M' coincide. We can express this by

$$L_{ij}(r_k,\ t) = \int R_{ij}(r_k)\ dr_k$$

which shows that scaling of turbulence (in tensorial form) depends on a vectorial quantity.

We do not intend to carry out a complete spectral analysis, but we note that an equation can be given for Φ_{ij} in spectral space, which corresponds to the equation for R_{ij} in physical space.

We mention only the main scaling processes here and consider only isotropic fields. Readers interested in more details are encouraged to consult specialized treatises.

In an isotropic turbulence we find that

$$\frac{\overline{q^2}}{2} = \frac{\overline{u_i u_i}}{2} = \int_0^\infty E(k)\ dk$$

where k stands for the modulus of **k**. $E(k)$ is the spectral distribution of the turbulent kinetic energy (related to Φ_{ij}):

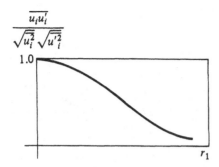

Fig. 4.5   Correlation function

The turbulent kinetic energy for an isotropic field cannot be fed from a mean motion, so the equation for energy balance becomes

$$\frac{d\overline{q^2}}{dt} = -2\overline{\varepsilon}$$

The dissipation term $\overline{\varepsilon}$ is linked to $E(k)$ by the relation

$$\overline{\varepsilon} = 2v \int_0^\infty k^2 E(k, t) \, dk \cong v \left[\frac{\partial u_i}{\partial x_j} \right]^2 \tag{4.16}$$

In this context, Kolmogoroff's assumptions regarding the spectrum of energy can be stated briefly:

1. A large amount of energy is stored in the part of the spectrum that corresponds to small values of k, that is, to large turbulent structures. This spectral range feeds another part of the spectrum through non-linear interactions.
2. An inertial range exists, in which an amount of energy is conveyed from the largest to the smallest structures. Viscosity effects can be neglected, so the energy flux is constant.
3. Viscous effects become significant at large values of k. In this part of the spectrum, mechanical energy is transformed into heat.

In this frame proposed by Kolmogoroff, each part of the spectrum plays an ad hoc role that assumes that long-range interactions are not significant. The spectrum does not exhibit peaks or valleys; the energy flux being constant, energy cannot be stored in a narrow band of the inertial range. Even though viscous effects are dominant in only a very limited part of the spectrum, their role is determinant because an asymptotic situation is imposed: Starting from any turbulent state, the fluid comes to rest after some delay. The concept of equilibrium implies that a strong connection should exist among all the parts of the spectrum (Fig. 4.6).

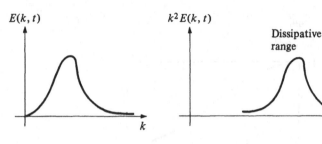

$E(k, t)$ $k^2 E(k, t)$ Dissipative range

Fig. 4.6 Spectra

It can be shown that the dissipation rate $\bar{\varepsilon}$ can be expressed as a function of u' and λ. The microscale λ is also the intercept on the time axis of the osculator parabola that matches the autocorrelation curve at the origin. The autocorrelation is defined as a function of the time difference $\tau = t' - t$. The microscale is a scale representative of the size of the smallest structures. Microscale and macroscale (defined with respect to either time or space) are represented in Fig. 4.7.

Introducing a global time T, we can set

$$\frac{d\overline{q^2}}{dt} = \frac{\overline{q^2}}{T}$$

If we take into account the coherence of the mechanism, we can introduce a characteristic length scale that is related to the time scale through a global parameter $\overline{q^2}$, as in the equation

$$T \approx \frac{L}{\sqrt{\overline{q^2}}}$$

which leads to

$$\frac{d\overline{q^2}}{dt} = \frac{[\overline{q^2}]^{3/2}}{L}$$

If comparison is made with

$$\frac{d\overline{q^2}}{dt} = \nu \frac{\overline{q^2}}{\lambda^2} \tag{4.17}$$

we obtain

$$\frac{\lambda}{L} \approx \frac{1}{\sqrt{R_L}} \qquad \text{with} \qquad R_L = \frac{L\sqrt{\overline{q^2}}}{\nu}$$

or

$$\frac{\lambda}{L} \approx \frac{1}{R_\lambda}$$

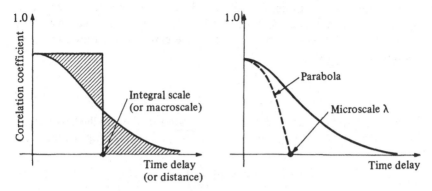

Fig. 4.7   Definitions of scales

One of the fundamental features of the spectrum concerns its shape. When an equilibrium state exists, the small turbulent structures adapt themselves to dissipate an amount of energy that is fixed by the dynamic properties of the largest structures. Accordingly, $\bar{\varepsilon}$ is a representative measure of this energy flux.

In the inertial zone of the spectrum the unique parameters that play any role are the energy flux $\bar{\varepsilon}$ through the zone and the modulus of the wave number k. A functional form embodying those variables can be written as

$$f\{E, k, \bar{\varepsilon}\} = 0 \qquad (4.18)$$

which, by dimensional analysis, gives

$$\frac{E}{\bar{\varepsilon}^{2/3} k^{-5/3}} = \text{constant}$$

In the viscous zone, the viscosity comes explicitly into play, so we can write

$$g\{E, k, \bar{\varepsilon}, \nu\} = 0$$

A similar dimensional analysis leads to

$$\frac{E}{\nu^{5/4}\bar{\varepsilon}^{1/4}} = f\left\{ k\left(\frac{\nu^3}{\bar{\varepsilon}}\right)^{1/4} \right\} = f(k\eta)$$

A new scale is introduced,

$$\eta = \left(\frac{\nu^3}{\bar{\varepsilon}}\right)^{1/4}$$

which is the Kolmogoroff length scale. This scale is characteristic of small structures because it depends on the unique quantities ν and $\bar{\varepsilon}$.

For dissipative structures, we can introduce a characteristic time τ_D that is based on both the size η of these structures and the kinematic viscosity coefficient ν. For small structures this time can be expressed on dimensional grounds as

$$\tau_D \approx \frac{\eta^2}{\nu}$$

A kinematic characteristic time τ_C, related to the smallest structures is given by the local velocity gradient $\overline{(\partial u_i/\partial x_j)^2}$, itself proportional to

$$\bar{\varepsilon} \approx \nu \left(\frac{\partial u_i}{\partial x_j} \right)^2 \tag{4.19}$$

from which we have

$$\tau_C \approx \left(\frac{\nu}{\bar{\varepsilon}} \right)^{1/2} \tag{4.20}$$

At this stage we assume the existence of local zones in the fluid (responsible for the dissipation) in which the two mechanisms are in competition, so the two time scales should be of the same order of magnitude. That leads to

$$\eta = \left(\frac{\nu^3}{\bar{\varepsilon}} \right)^{1/4} \tag{4.21}$$

The physical meaning of Kolmogoroff's scale is thereby displayed. A diffusive time related to the boundary layer thickness is introduced, and an advective time related to both a running length and the external velocity is required.

With a view to introducing chemical problems, we need to introduce a similar scaling for scalar quantities. The Prandtl and the Lewis numbers must account for the diffusing of scalar quantities such as temperature and chemical species. For laminar flows, they intervene in the development of the whole flow; whereas for turbulent flows, the mixing processes are controlled by the large turbulent structures except over a limited range of the spectrum where molecular effects come into play.

This similarity between the approach (that carried out at the scale of the whole flow and that carried out at the scale of small structures) can be expanded. For example, $\Sigma \, \partial u_i/\partial x_j$ is related to an expansion of a volume if the suffix j goes from one to three, and to an expansion of a surface if it goes from one to two. We can define a local expansion of a surface by introducing

$$\left[\sum_j \overline{\left(\frac{\partial u_i}{\partial x_j} \right)^2} \right]^{1/2} \qquad j \in 1, 2$$

This parameter may be relevant to determining the local probability of extinction of a turbulent flame. When the parameter is large, a local equilibrium between advection diffusion and heat release cannot exist, the characteristic time of chemical reaction being of the same order of magnitude as those associated with local gradients.

Because the straining is controlled by terms such as $\partial U_i/\partial x_j + \partial U_j/\partial x_i$, it would be interesting to introduce turbulent parameters such as

$$\left\{ \overline{\left(\frac{\partial U_i}{\partial x_j} + \frac{\partial U_j}{\partial x_i} \right)^2} \right\}^{1/2}$$

which would be closely related to the local straining of a flame front in a turbulent field. In other words, mean quantities related to typical features can suggest the introduction of turbulent parameters having the same meaning. A statistical point of view must be introduced.

4.3.b   *Scalar fields embedded in kinematic fields*

A complete treatment of this problem is long and we outline only the most important features in this section: We first examine the main relations between the kinematic fluctuating field **U** and the scalar field *C*. This discussion is fundamental for chemical processes.

The instantaneous values of the two fields are linked through the equation

$$\frac{\partial C}{\partial t} + U_j \frac{\partial C}{\partial x_j} = D \frac{\partial^2 C}{\partial x_j \, \partial x_j} \qquad (4.22)$$

The diffusion of a scalar quantity is controlled by the coefficient *D*. If the whole field is split into a mean field and a turbulent field, the correlations $\overline{u_j c}$ appear in the equation for the evolution of $\overline{C}$. The spectrum distribution of the scalar quantity is denoted $E_c(k, t)$ and the field is assumed to be isotropic:

$$\frac{\overline{c^2}}{2} = \int_0^\infty E_c(k, t) \, dk \qquad (4.23)$$

The scalar fluctuations are destroyed by a diffusion process called $\overline{\varepsilon_c}$, which plays a role similar to that of the viscous effects in the kinematic field previously examined. The corresponding decay of the fluctuation field can be expressed as

$$\frac{\partial \overline{c^2}}{\partial t} = -2\overline{\varepsilon}_c$$

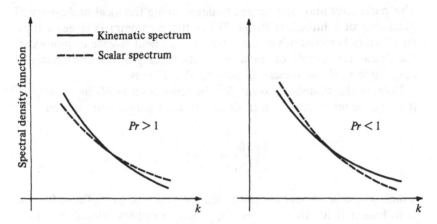

Fig. 4.8   Influence of the Prandtl number on the shape of the spectrum

with

$$\bar{\varepsilon}_c = 2D \int_0^\infty k^2 E_c(k, t)dk \qquad (4.24)$$

The Prandtl or Lewis numbers result from a comparison between viscous and diffusive effects. The relative position of the two spectral curves can be surmised from the value of the ratio v/D (Fig. 4.8).

Over a wide range of wavenumbers, the spectrum equilibrium of the scalar field is a direct consequence of the spectrum equilibrium of the kinematic field. The largest structures are strongly marked by the mechanisms that generate turbulence, and the inertial range is dominated by inertial exchanges with constant fluxes of energy or matter, whereas the range of the small structures is dominated by typical diffusive effects.

With the following assumptions, we can seek universal laws for each range of the spectrum. These ranges are the energy containing (I), the inertial (II), the viscoadvective (III), and the viscodiffusive (IV), as shown in Fig. 4.9.

> The two turbulences are generated by the same system (e.g., a passive grid).
> The Reynolds number is sufficiently large to permit the existence of an inertial range.
> The Lewis (or Prandtl) number is significantly different from 1, to clearly emphasize the two phenomena. $(Pr \gg 1)$.

In the inertial range (zone II), inertial effects predominate, and the energy flux is assumed to be constant for both phenomena. A universal law is then postulated in the form proposed by Corrsin:

$$f\{E_c, k, \bar{\varepsilon}, \bar{\varepsilon}_c\} = 0$$

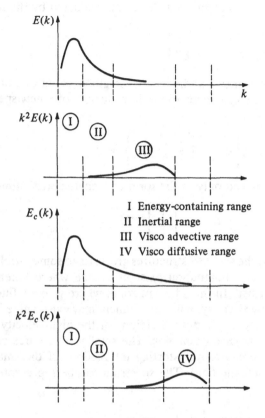

Fig. 4.9   The classical ranges of the turbulence spectra

The scalar cascade is characterized by $\bar{\varepsilon}_c$. Four quantities appear that depend on three fundamental units. A unique dimensionless quantity can be introduced. That leads to

$$\frac{E_c}{\bar{\varepsilon}^{-1/3}\bar{\varepsilon}_c} \approx k^{-5/3} \tag{4.25}$$

E_c appears to be proportional to ε_c, which can be considered a consequence of the equation for C. This equation is linear with respect to C if the kinematic field is assumed to be unaltered. Experimental data seem to support this law.

For zone III, we reproduce an analysis proposed by Batchelor.

$$g\{E_c,\ k,\ v,\ \bar{\varepsilon},\ \bar{\varepsilon}_c\} = 0$$

Because the action of the kinematic field on the scalar field is dominated by local mean velocity gradients, which can be evaluated by the term

$$\left[\frac{\overline{\varepsilon}}{\nu}\right]^{1/2}$$

we can conjecture that these gradients are the global result of a double contribution $\overline{\varepsilon}$ and ν, which leads to the introduction of a new spectral function:

$$g\left\{E_c,\ k,\ \left[\frac{\nu}{\overline{\varepsilon}}\right]^{1/2},\ \overline{\varepsilon}_c\right\} = 0$$

$\overline{\varepsilon}$ and ν can be introduced only in the form of a characteristic time:

$$E_c \approx \varepsilon_c \left[\frac{\nu}{\overline{\varepsilon}}\right] k^{-1}$$

The equilibrium of the smallest structures gives rise to some problems. For instance, can scalar fluctuation exist in a zone where kinematic fluctuations do not exist? In the adjacent zone, where velocity fluctuations are destroyed by viscosity, a limited domain may exist where high-strain processes reflect the combined actions of the high-velocity gradients located in the adjacent domains. The smallest structures of the scalar field are also created by distorting interactions of the smallest structures of the kinematic field. The spectrum can be represented in the form

$$h\left\{E_c,\ k,\ \left[\frac{\nu}{\overline{\varepsilon}}\right]^{1/2},\ \overline{\varepsilon}_c,\ D\right\} = 0 \qquad (4.26)$$

The comparison between the two scales η and η_c that indicate the characteristic sizes of the structures subjected to viscous and diffusive effects, respectively, is important. These two scales are given by the following:

kinematic field $\tau_c = \left[\dfrac{\nu}{\overline{\varepsilon}}\right]^{1/2}$ $\tau_D = \dfrac{\eta^2}{\nu} \Rightarrow \eta \approx \left[\dfrac{\nu^3}{\overline{\varepsilon}}\right]^{1/4}$

scalar field $\tau_c = \left[\dfrac{\nu}{\overline{\varepsilon}}\right]^{1/2}$ $\tau_D^S = \dfrac{\eta_c^2}{\nu} \Rightarrow \eta_c \approx \left[\dfrac{D^2\nu}{\overline{\varepsilon}}\right]^{1/4}$

So the spectrum shape finally becomes

$$\frac{E_c\,\nu}{\eta^3\varepsilon_c[D/\nu]^{1/2}} = g(k\eta_c)$$

4.4 GLOBAL BEHAVIOR OF TURBULENT BOUNDARY LAYERS

4.4.a *Kinematic boundary layers*

Turbulent flows occur more frequently in nature than laminar flows, but they cannot be analyzed satisfactorily from a theoretical standpoint. In the past, relationships for predicting mean velocities have been derived using Prandtl's mixing-length theories as a basis. In these theories (as in Boussinesq's approach) a local dependence is postulated between the mean velocity profile and the shear stress, that is, between the mean velocity profile and the turbulent fluid ($\overline{uv}$). Such a connection cannot be considered valid. Statistical turbulence quantities are governed by differential equations that determine their evolution with respect to time; accordingly, these quantities exhibit memory effects. The correlation coefficient can be shown to depend on the mean velocity gradient $\partial \overline{U_1}/\partial x_2 \equiv \partial \overline{U}/\partial y$ applied in a cumulative way. In other words, $\overline{uv}/\sqrt{\overline{u^2}}\sqrt{\overline{v^2}}$ depends on $\int(\partial \overline{U_1}/\partial x_2)\,dt$ and not on the local value of this gradient $\partial \overline{U_1}/\partial x_2$. In fact, Prandtl's or Boussinesq's theories were first applied to simple flows in which equilibrium turbulent states exist. In channel flows far from the exit, the generation of turbulent structures is marked by both the shape of the mean velocity profile and by the mean shear stress distribution through the production term $\overline{uv}(\partial \overline{U}/\partial y)$. Moreover, these structures follow a mean path located at a given distance from the wall. For these two reasons an equilibrium state exists. The turbulent structures are oriented by a constant mean velocity field and randomized by the turbulent field, whose statistical properties are independent of the x coordinate (Fig. 4.10).

In many flows a slow evolution exists that does not obliterate a quasi-equilibrium state, so no evident failure can be detected. This failure appears if the flow is subjected to rapid distortion effects, which often result from either a rapid change of the frontier (distorting duct) or a mixing process (two flows exhibiting typical properties merge into each other, Fig. 4.11).

Memory effects come into play in all of these cases, and a local connection between the statistical properties of the turbulent field and the local mean velocity profile does not exist. To bridge these gaps more elaborate theories have been developed that take memory effects into account in a more or less refined way. The statistical equation of turbulence seems sufficient in general to introduce the equation relative to $\overline{u_1 u_2}$ and to postulate closure assumptions at this level.

We shall give elementary information concerning these prediction methods based on a dimensional analysis supported by qualitative re-

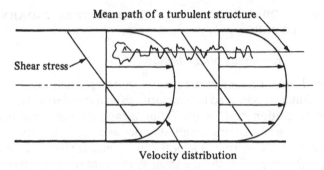

Fig. 4.10   Turbulent flow through a pipe

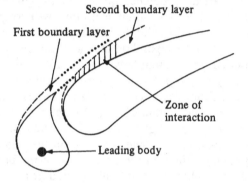

Fig. 4.11   Boundary layers in cascade

marks. The relevant parameters that intervene in wall-bounded shear flows are the kinematic viscosity v, the wall shear stress τ_p, the undisturbed velocity $\bar{u}_m$, and the boundary layer thickness δ. Furthermore, if we assume the mean velocity $\bar{U}$ at the y coordinate to depend on all these parameters, we can write

$$\bar{U} = f\{y, v, \tau_p, \bar{u}_m, \delta\}$$

At this stage, we replace the wall shear stress τ_p by a fictitious velocity u_f, which is termed *friction velocity* and is defined by the relation

$$u_f^2 = \frac{\tau_p}{\rho}$$

So far, six quantities have been introduced that depend on two fundamental dimensions, so an adequate dimensional analysis requires only four dimensionless parameters. This state can be expressed as

$$\frac{\bar{U}}{u_f} = \varphi\left\{\frac{yu_f}{v}, \frac{y}{\delta}, \frac{\bar{u}_m}{u_f}\right\}$$

Turbulence production is locally the consequence of an interaction between the mean velocity gradient and the turbulent velocity correlation:

$$\frac{\partial \overline{U}}{\partial y} \, \overline{uv}$$

According to experimental data already mentioned, a large amount of the turbulent kinetic energy is produced next to the wall in a region where both the mean velocity gradient and the correlation term are very large. Interactions between the inner and outer parts of the boundary layer can be expected; however, it is statistically evident that these exchanges are dominated by information that goes from the wall region to the outer region. These mechanisms are a one-way street to a large extent. The amount of energy conveyed from the wall region to the outer zone is rather small, so the wall region can be considered to be somewhat independent of the overall flow. This translates into the following equation:

$$\text{turbulence production} = \text{dissipation}$$

indicating that the advective and diffusive terms are small compared with the production terms.

In the wall region, the mean velocity should depend on y, ν, and u_f only.

$$\overline{U} = f\{y, \, \nu, \, u_f\}$$

$$\frac{\overline{U}}{u_f} = f\left\{\frac{yu_f}{\nu}\right\}$$

This region is traditionally split into two zones: the *viscous sublayer* and the *wall turbulent region*. In the viscous sublayer, the contribution of turbulence to shear stress effects is negligible, so we can write

$$\tau_p = \mu \frac{d\overline{U}}{dy}$$

$$\frac{\tau_p}{\rho} = u_f^2 = \nu \frac{d\overline{U}}{dy}$$

and by integration

$$\frac{\overline{U}}{u_f} = \frac{yu_f}{\nu}$$

The wall turbulent region is characterized by the dominant role of u_f and ν. The kinematic viscosity does not contribute directly to the shear stress because the mean velocity gradient becomes small in this region

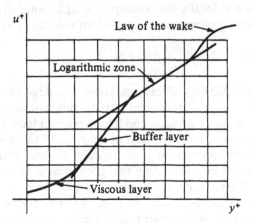

Fig. 4.12   The various laws of the boundary layer

and also because the contribution of turbulence to the shear stress becomes very large through the $\overline{uv}$ terms.

From experimental data a "law of the wall" can be proposed:

$$\frac{\overline{U}}{u_f} = A + B \log \frac{y u_f}{v}$$

Viscous effects in the outer part of the boundary layer are assumed not to interfere with the mean velocity profile, which is defined with reference to the maximum mean velocity u_m,* and is expressed in the final form of a velocity defect law:

$$\overline{U} = \varphi\{y, \, \delta, \, u_f, \, \overline{u_m}\}$$

$$\frac{U}{u_f} = \varphi\left\{\frac{y}{\delta}, \, \frac{u_m}{u_f}\right\}$$

$$\frac{\overline{U} - \overline{u_m}}{u_f} = f_i\left\{\frac{y}{\delta}\right\}$$

In Fig. 4.12, $\overline{U}/u_f$ is termed u^+ and $y u_f/v$ is denoted by y^+. Moreover, the existence of a layer between the viscous sublayer and the logarithmic region is presented as a buffer layer.

Boussinesq introduced a turbulent viscosity coefficient v_T to extend the simple relationship encountered for laminar flows to the case of

---

* From this external point of view, all the regions in which viscous effects play a role are removed from this analysis. The "law of the wake" is available regarding the outer part of the boundary layer.

Table I. *The regions of a boundary layer.*

Region	$y^+ = \dfrac{yu_f}{\nu}$	Law $u^+ = \dfrac{u}{u_f}$
Viscous sublayer	0–7	$u^+ = y^+$
Buffer layer	7–30	$u^+ = 3.05 + 11.5 \log y^+$
Logarithmic friction law		
(law of the wall)	30–400	$u^+ = 5.5 + 5.75 \log y^+$
Central part		$\nu_T = $ constant
(law of the wake)	400–500	A parabolic profile exists with a fictitious viscosity

turbulent flows. A first gradient approximation can be written as

$$\tau = \nu_T \frac{d\overline{U}}{dy} \tag{4.27}$$

where ν_T is approximately constant in the outer region of the boundary layer. The law of the wall describes the inner turbulent part of the boundary layer whereas the law of the wake applies to the outer turbulent part of the boundary layer. If an overlapping region exists in which the two laws are satisfied, a logarithmic function can be introduced,

$$\frac{\overline{U}}{u_f} = A + B \log \frac{yu_f}{\nu} \tag{4.28}$$

in which the constants such as ν can be removed by subtraction. Typical values of the coefficients for the various regions of the boundary layer are presented in Table I.

The main parameters that can act on the velocity distribution can be classified as external or internal.

External parameters: The presence of an adverse pressure gradient has the effect of reducing the logarithmic zone and reinforces the wake region, as shown in Fig. 4.13.

As previously mentioned, the shear stress distribution is altered. The maximum shear stress is located at a distance y, which increases from the wall as the adverse pressure gradient grows. The turbulence production is not as localized so the influence of the wall decreases up to the separation point where the velocity changes sense.

Where a highly turbulent external field exists, such as in a wall jet, two production centers are in competition; the first is located next to the wall and the second in the jet region, so two kinds of turbulence are in competition in a very extended domain. Figure 4.14 refers to the

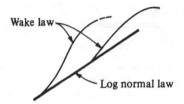

Fig. 4.13   Influence of an adverse pressure gradient

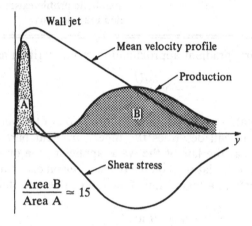

Fig. 4.14   Wall jet

small region between the wall and the ordinate δ where the velocity is maximum. Turbulent exchanges strongly increase, so the coherence of the flow also increases. However, the region next to the wall is not significantly affected, and consequently the viscous sublayer and the buffer layer are not modified.

Internal parameters: Two internal factors that can affect the velocity distribution are the wall roughness, which introduces a new scaling linked to the sizes of the roughness, and an injection (or suction) at the wall. These two actions, which develop in the wall region, are capable of significantly disturbing the flow next to the wall (Fig. 4.15).

4.4.b   *Mean properties of a thermal boundary layer*

A similar approach can be carried out using dimensional analysis and Boussinesq's concept. First, we assume the shear stress to be propor-

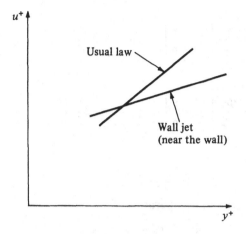

Fig. 4.15   Influence of an external field on the log law

tional to the mean velocity field by a first-gradient approximation of the type

$$-\overline{uv} = v_{\text{T}} \frac{d\overline{U}}{dy} \qquad (4.29)$$

Thus, the total shear stress can be expressed as

$$\frac{\tau}{\rho} = [v + v_{\text{T}}] \frac{d\overline{U}}{dy}$$

The heat flux can be written as

$$\Phi = \lambda \frac{d\overline{T}}{dy} - \rho C \overline{v\theta}$$

A first-gradient approximation yields

$$-\overline{v\theta} = v_{\text{T}}^{\text{H}} \frac{d\overline{T}}{dy}$$

and we find that

$$\frac{\Phi}{\rho C} = \left[\frac{\lambda}{\rho C} + v_{\text{T}}^{\text{H}} \right] \frac{d\overline{T}}{dy} = \left[\frac{v}{Pr} + v_{\text{T}}^{\text{H}} \right] \frac{dT}{dy} \qquad (4.30)$$

Next, the two coefficients v_{T} and v_{T}^{H} are assumed to be the same, stipulated by Reynolds analogy.

If both the shear-stress and the heat-flux distributions are assumed to be fairly well represented by a linear law, we have to compare the two algebraic relations

$$\frac{\tau}{\rho} = \frac{\tau_w}{\rho}\left[1 - \frac{y}{\delta}\right] = (\nu + \nu_T)\frac{d\bar{U}}{dy}$$

and

$$\frac{\Phi}{\rho C} = \frac{\Phi_w}{\rho C}\left[1 - \frac{y}{\delta}\right] = \left(\frac{\nu}{Pr} + \nu_T^H\right)\frac{d\bar{T}}{dy} \tag{4.31}$$

The friction velocity

$$\tau_w = \rho u_f^2$$

is associated with the friction temperature T_f

$$\Phi_w = \rho C u_f T_f$$

If we set

$$u^+ = \frac{\bar{U}}{u_f} \qquad T^+ = \frac{\bar{T}}{T_f}$$

we finally have

$$1 - \frac{y}{\delta} = \frac{\nu + \nu_T}{u_f}\frac{du^+}{dy}$$

$$1 - \frac{y}{\delta} = \frac{\nu/Pr + \nu_T}{u_f}\frac{dT^+}{dy}$$

If the Prandtl number is equal to 1, it follows from the previous relations that

$$du^+ = dT^+ \quad \text{and} \quad \frac{\bar{u}_m - \bar{u}_w}{u_f} = \frac{\bar{T}_m - \bar{T}_w}{T_f}$$

The no-slip condition demands that $U_w = 0$, so

$$T_f = \frac{u_f}{u_m}[\bar{T}_m - \bar{T}_w]$$

$$\Phi = \rho C u_f^2\frac{[\bar{T}_m - \bar{T}_w]}{\bar{u}_m} = \frac{C_f}{2}\rho C\bar{u}_m[\bar{T}_m - \bar{T}_w] \tag{4.32}$$

Comparing this heat flux with an advective flux based on u_m and $T_m - T_w$, we can write

$$Ma = \frac{\Phi_w}{\rho C u_m[T_m - T_w]} = \frac{C_f}{2}$$

in which *Ma* is the Margoulis number.

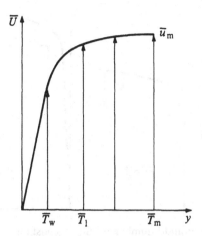

Fig. 4.16   Thermal boundary layer

If the Prandtl number is different from unity, integration is performed over two separate areas. The first integration is performed between the wall and the outer edge of the viscous layer, in the region where the Prandtl number plays an important role, and the second integration is performed in the fully developed turbulent zone with the same assumption

$$\nu_T = \nu_T^H$$

In the sublayer where $\nu \gg \nu_T$, we have

$$Pr\, du^+ = dT^+$$

$$Pr\, \frac{\overline{U}}{u_f} = \frac{\overline{T}_1 - \overline{T}_w}{T_f} \tag{4.33}$$

and in the turbulent zone, by assuming

$$\nu_T = \nu_T^H \qquad \text{and} \qquad \nu_T \gg \nu$$

we obtain

$$du^+ = dT^+$$

$$\frac{\overline{u}_m - \overline{u}_1}{u_f} = \frac{\overline{T}_m - \overline{T}_1}{T_f} \tag{4.34}$$

Eliminating T_1 between Eq. 4.33 and Eq. 4.34 we have

$$Ma = \frac{C_f}{2} \frac{1}{1 - u_1/u_m(1 - Pr)}$$

A thermal barrier is consequently erected through the sublayer next to the wall (Fig. 4.16). Whatever the turbulence level in the outer part of

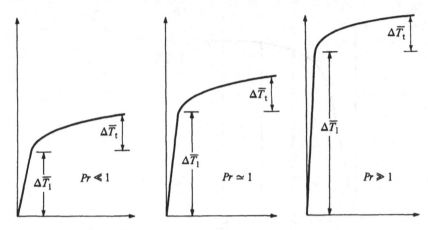

Fig. 4.17   Role of the Prandtl number for Φ_w = constant

the boundary layer, conduction effects prevail. Thermal exchanges can be accelerated only by an appropriate choice of fluid. The only requirement for the Prandtl number is that it be high. Figure 4.17 shows the role of the Prandtl number when the heat flux is constant.

When Φ_w is kept constant, the difference of temperature in the turbulent zone is also constant. Differences of temperature in the laminar zone are strongly dependent on the Prandtl number, which explains the choice of liquid metal to enhance thermal exchanges.

$$\Delta T = \underbrace{\Delta T_t - \Delta T_l}_{\text{this term depends on } Pr}$$

Some refinements have been made that do not alter the basis of this computation. Only the splitting of the wall region is altered—for instance, the region of the buffer layer, which is treated independently.

We must discuss the role of the boundary conditions briefly. In the previous approach, in which kinematic and thermal boundary conditions were assumed to be similar, the consequence is that the heat flux is null at the point where the shear stress is null. In many industrial cases, however, thermal and kinematic boundary conditions may be independent of each other. The heat flux is modulated by the thermal properties of the wall because these properties have no significant effects on the flow (Fig. 4.18).

Computations can be carried out if we assume the heat flux and the shear stress are null at the location δ' and Δ', respectively. The two relations can be expressed as

$$1 - \frac{y}{\delta'} = \frac{\nu + \nu_T}{u_f} \frac{d\overline{U}}{dy}$$

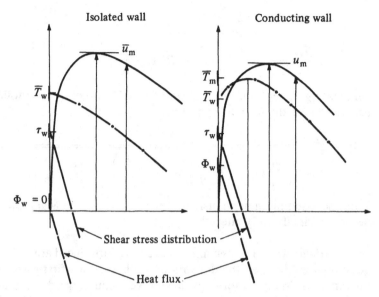

Fig. 4.18   Case of the wall jet

$$1 - \frac{y}{\Delta'} = \frac{\nu/Pr + \nu_T^H}{u_f}\frac{d\overline{T}}{dy}$$

Noting that in the sublayer the two quantities y/δ' and y/Δ' are very small, we find that

$$Pr\, du^+ = dT^+ \quad \text{and} \quad Pr\frac{\overline{u}_1}{\overline{u}_m} = \frac{\overline{T}_1 - \overline{T}_w}{\overline{T}_f} \qquad (4.35)$$

In the turbulent zone, ν_T^H is still assumed to be equal to ν_T, so

$$\begin{cases} 1 = \dfrac{\nu_T}{u_f}\dfrac{1}{1-y/\delta'}\dfrac{du^+}{dy} \\ 1 = \dfrac{\nu_T}{u_f}\dfrac{1}{1-y/\Delta'}\dfrac{dT^+}{dy} \end{cases}$$

Therefore, we find that

$$dT^+ = du^+ + \left[\frac{1}{\delta'} - \frac{1}{\Delta'}\right]\frac{y}{1-y/\delta'}du^+ \qquad (4.36)$$

If the crucial region for thermal exchanges is the logarithmic region (not the parabolic zone), we can write

$$u^+ = A + B\log\frac{yu_f}{\nu}$$

$$du^+ = B \frac{dy}{y}$$

$$\frac{\overline{T}_m - \overline{T}_1}{T_f} \cong \frac{\overline{u}_m - \overline{u}_1}{u_f} - B\left[1 - \frac{\delta'}{\Delta'}\right] \log\left|1 - \frac{\Delta'}{\delta'}\right| \qquad (4.37)$$

Eliminating T_1 between Eq. 4.35 and Eq. 4.37, we obtain the following expression for the Margoulis number:

$$Ma = \frac{C_f}{2} \frac{1}{1 - \dfrac{u_1}{u_m}(1 - Pr)\sqrt{\dfrac{C_f}{2} B\left(1 - \dfrac{\delta'}{\Delta'}\right)\log\left|1 - \dfrac{\Delta'}{\delta'}\right|}} \qquad (4.38)$$

With the assumption previously set forth, $v_T^H = v_T$, the Margoulis number is not significantly altered.

Remarks: The assumptions we have introduced are in fact very questionable. First, all the derivations are based on a first-gradient approximation. Many patterns of flows reveal countergradient heat flux or countergradient momentum flux. A first-gradient approximation is fairly useful when an equilibrium state exists. In highly asymmetrical flows, such as wall jets and channel flows with smooth and rough walls, a countergradient flux is observed. In fact, these regions of countergradient flux are limited in extent, and accordingly, the previous relations are of interest for engineers. Second, the use of a same coefficient to link the momentum to the gradient of mean velocity and to link the heat flux to the gradient of mean temperature is also questionable, the velocity gradient being a second-order tensor whereas the temperature gradient is a first-order tensor (vector) as expressed by

$$\overline{u_i u_j} = v_T \frac{\partial \overline{u}_i}{\partial x_j} \qquad \overline{u_i \theta} = v_T^H \frac{\partial \overline{T}}{\partial x_i}$$

Perhaps, as already proposed, it is better to establish an analogy using the turbulent kinetic energy in place of $\overline{u_i u_j}$.

4.5   THE CONCEPT OF BOUNDARY LAYER IN TURBULENT FLOWS: CASE OF THE FREE JET

We can carry out an analysis of the order of magnitude of all the terms that intervene in the equation of a jet like we previously did for laminar flows, even though the evaluation of the order of magnitude of mean turbulent quantities is somewhat questionable.

For most (nearly two-dimensional) flows, we need two length scales and two velocity scales; the length scales are those in the longitudinal

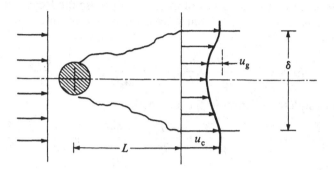

Fig. 4.19 Wake definition sketch

(L) and the transverse (δ) direction whereas one of the velocity scales represents convection (u_c) and one is important for gradient properties (u_g) such as stress. See Fig. 4.19.

For boundary layers and jets (with zero velocity outside the jet), we only need L, δ, and $u_c = u_g$ whereas for wakes we need all four scales; in an even simpler flow in pipes, we need only δ and u_c (because the flow is independent of L). Boundary layers and jets are called *semi-constrained* flows whereas wakes are called *unconstrained* flows.

Consider boundary layers and/or jets for which there is only one velocity scale; if $\overline{U} \approx u_g$, then for the terms in the continuity equation to balance, we find for V

$$\frac{\partial \overline{U}}{\partial x} \approx \frac{u_g}{L} \qquad \frac{\partial \overline{V}}{\partial y} \approx \frac{\overline{V}}{\delta} \qquad \frac{\partial \overline{U}}{\partial x} + \frac{\partial \overline{V}}{\partial y} = 0 \Rightarrow \overline{V} \approx u_g \frac{\delta}{L}$$

Similarly for the two Navier–Stokes equations,

$$\overline{U}\frac{\partial \overline{U}}{\partial x} \approx \frac{u_g}{L} \qquad \overline{V}\frac{\partial \overline{U}}{\partial y} \approx U_g \frac{\delta}{L}\frac{u_g}{\delta} = \frac{u_g}{L}$$

as well; from this we see that the Reynolds stress term must be of the order of u_j^2/L, which can be expressed as

$$\frac{\partial(\overline{uv})}{\partial y} \approx \frac{u_g^2}{L} \Rightarrow \frac{\overline{u'^2}}{\delta} \approx \frac{u_g^2}{L}$$

Because $u' \approx v'$, we find that

$$\mu'^2 \approx \frac{\delta}{L} u_g^2 \qquad \mu'^2 = \overline{\mu^2}$$

Furthermore, from $-\rho\overline{uv} = \mu\,\partial\overline{U}/\partial y$, we can find the order of the molecular viscosity. (Certainly the molecular viscosity is overestimated because to determine its order of magnitude we use the Reynolds stress $\overline{uv}$; we should keep in mind that viscosity is important in the vicinity of solid boundaries.)

$$\overline{uv} \approx u'^2 \qquad \frac{\partial \overline{U}}{\partial y} = \frac{u_g}{\delta} \Rightarrow v \approx \frac{\delta u'^2}{u_g} \qquad \text{or} \qquad \frac{\delta^2 u_g}{L}$$

So the complete equation in the x direction is

$$\overline{U}\frac{\partial \overline{U}}{\partial x} + \overline{V}\frac{\partial \overline{U}}{\partial y} = -\frac{1}{\rho}\frac{\partial \overline{P}}{\partial x} - \frac{\partial(\overline{uv})}{\partial y} - \frac{\partial \overline{u^2}}{\partial x} + \nu\frac{\partial^2 \overline{U}}{\partial x^2} + \nu\frac{\partial^2 \overline{U}}{\partial y^2}$$

$$\downarrow \qquad\qquad \downarrow \qquad\qquad\qquad\qquad \downarrow \qquad\qquad \downarrow \qquad\qquad \downarrow \qquad\qquad \downarrow$$

$$\frac{u_g^2}{L} \qquad \frac{u_g^2}{L} \qquad\qquad\qquad \frac{u_g^2}{L} \qquad \left(\frac{\delta}{L}\right)\frac{u_g^2}{L} \qquad \left(\frac{\delta}{L}\right)\frac{u_g^2}{L} \qquad \frac{u_g^2}{L}$$

and the corresponding equation in the y direction is

$$\overline{U}\frac{\partial \overline{V}}{\partial x} + \overline{V}\frac{\partial \overline{V}}{\partial y} = -\frac{1}{\rho}\frac{\partial \overline{P}}{\partial y} - \frac{\partial(\overline{uv})}{\partial x} - \frac{\partial \overline{v^2}}{\partial x} + \nu\frac{\partial^2 \overline{V}}{\partial x^2} + \nu\frac{\partial^2 \overline{V}}{\partial y^2}$$

$$\downarrow \qquad\qquad \downarrow \qquad\qquad\qquad\qquad \downarrow \qquad\qquad \downarrow \qquad\qquad \downarrow \qquad\qquad \downarrow \qquad (4.39)$$

$$\left(\frac{\delta}{L}\right)\frac{u_g^2}{L} \quad \left(\frac{\delta}{L}\right)\frac{u_g^2}{L} \qquad\qquad \left(\frac{\delta}{L}\right)\frac{U_g^2}{L} \quad \frac{u_g^2}{L} \quad \left(\frac{\delta}{L}\right)^3\frac{u_g^2}{L} \quad \left(\frac{\delta}{L}\right)\frac{u_g^2}{L}$$

So, for $\delta/L \to 0$ (or to zero order in δ/L), the second term on the right-hand side survives, as well as the pressure term (of which we know nothing); therefore,

$$-\frac{1}{\rho}\frac{\partial \overline{P}}{\partial y} - \frac{\partial \overline{v^2}}{\partial y} = O\left(\left(\frac{\delta}{L}\right)\frac{u_g^2}{L}\right)$$

Integrating from a point within the boundary layer or jet (see Fig. 4.20) to a point in the freestream, and noting that in the freestream $v_0^2 = 0$ and $p_0 = p_{(x)}$ only, we get

$$\overline{P} + \rho\overline{v^2} = P_0 + O\left(\delta\left(\frac{\delta}{L}\right)\frac{u_g^2}{L}\right)$$

This result is now substituted in the streamwise equation. In general, we neglect the resulting term, which corresponds to normal stress $(\partial\overline{v^2}/\partial x - \partial\overline{u^2}/\partial x)$ because we have assumed that $u \sim v$ (observed experimentally too). In any case, it is a $O(\delta/L)$ term, which is negligible for jets because there are no solid boundaries and viscous effects are negligible everywhere. To zeroth order we can then write

$$\overline{U}\frac{\partial \overline{U}}{\partial x} + \overline{V}\frac{\partial \overline{U}}{\partial y} = -\frac{1}{\rho}\frac{\partial P_0}{\partial x} - \frac{\partial(\overline{uv})}{\partial y}$$

We now examine the concept of similarity.

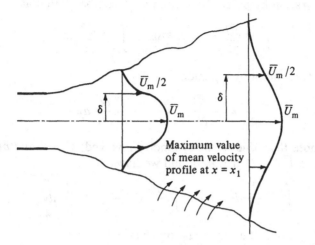

Fig. 4.20 Jet definition sketch

As previously shown, there are many ways of defining δ; here it is defined as the vertical distance between the centerline and the point at which the mean velocity is half the maximum at a given x downstream from the jet; similarity is postulated by assuming a velocity profile of the form

$$\overline{U}(x, y) = \overline{U}_M(x)f(\eta)$$

where $\eta = y/\delta$. Recall that the equation for a jet is

$$\overline{U}\frac{\partial\overline{U}}{\partial x} + \overline{V}\frac{\partial\overline{U}}{\partial y} = -\frac{1}{\rho}\frac{\partial P_0}{\partial x} - \frac{\partial(\overline{uv})}{\partial y}$$

but usually in the freestream outside the jet $P_0 = P_{atm}$, so $\partial P_0/\partial x = 0$, and we can write

$$\overline{U}\frac{\partial\overline{U}}{\partial x} + \overline{V}\frac{\partial\overline{U}}{\partial y} = -\frac{\partial(\overline{uv})}{\partial y}$$

In this form, advective mechanisms are controlled by the mean velocity whereas the diffusion process along the axis is dominated by the $\overline{uv}$ component of the Reynolds stress tensor.

The x axis is clearly an axis of symmetry. So we can integrate the last equation from 0 to y to yield

$$\int_0^y \overline{U}\frac{\partial\overline{U}}{\partial x} dy + \int_0^y \overline{V}\frac{\partial\overline{U}}{\partial y} dy = -(\overline{uv})|_0^y$$

After integration by parts, the second term can be written as

$$\int_0^y \overline{V} \frac{\partial \overline{U}}{\partial y} \, dy = (\overline{UV})|_0^y - \int_0^y \overline{U} \frac{\partial \overline{V}}{\partial y} \, dy$$

and from the continuity equation,

$$\frac{\partial \overline{V}}{\partial y} = -\frac{\partial \overline{U}}{\partial x} = \overline{V} = -\int_0^y \frac{\partial \overline{U}}{\partial x} \, dy$$

Now we note that, due to the symmetry involved, $V = 0$ and $\overline{uv} = 0$ at $y = 0$, so $\overline{UV} = 0$ at $y = 0$, and we have

$$\int_0^y \overline{V} \frac{\partial \overline{U}}{\partial y} \, dy = -U \int_0^y \frac{\partial U}{\partial x} \, dy + \int_0^y U \frac{\partial U}{\partial x} \, dy$$

Substituting this in the first equation gives

$$-\overline{uv} = 2 \int_0^y U \frac{\partial U}{\partial x} \, dy - U \int_0^y \frac{\partial U}{\partial x} \, dy \qquad (4.40)$$

or letting $y = \eta \delta$ gives $dy = \delta \, d\eta$; it has been assumed that $U = u_m f(\eta)$, so keeping in mind that $U_M = U_M(x)$,

$$2 \int_0^y \overline{U} \frac{\partial \overline{U}}{\partial x} \, dy = 2 \int_0^\eta \overline{U}_M f \frac{\partial}{\partial x} (\overline{U}_M f) \delta \, d\eta$$

$$= 2\delta U_M \int_0^\eta f \left(f \frac{dU_M}{dx} + U_M \frac{df}{dx} \right) d\eta$$

with

$$\frac{d\overline{U}_M}{dx} = \overline{U}'_M$$

and

$$\frac{\partial f}{\partial x} = \frac{\partial \eta}{\partial x} \frac{\partial f}{\partial \eta} \qquad \frac{\partial \eta}{\partial x} = \frac{\partial}{\partial x} \left(\frac{y}{\delta} \right) = -\frac{y}{\delta^2} \frac{d\delta}{dx} = -\frac{\eta}{\delta} \delta'$$

we have

$$2 \int_0^y \overline{U} \frac{\partial \overline{U}}{\partial x} \, dy = 2\delta \overline{U}_M \int_0^\eta f \left(f \overline{U}'_M - U_M \frac{\eta}{\delta} \delta' \frac{df}{d\eta} \right) d\eta$$

$$= 2\delta \overline{U}_M \overline{U}'_M \int_0^\eta f^2 \, d\eta - 2\delta' \overline{U}_M^2 \int_0^\eta \eta f \frac{df}{d\eta} \, d\eta$$

Also,

$$\overline{U}\int_0^y \frac{\partial \overline{U}}{\partial x}\, dy = \overline{U}_M f \int_0^{\eta}\left(f\overline{U}'_M - \overline{U}_M\frac{\eta}{\delta}\,\delta'\,\frac{df}{d\eta}\right)\delta\, d\eta$$

$$= \delta\overline{U}_M\overline{U}'_M f\int_0^{\eta} f\, d\eta - \delta'\overline{U}_M^2 f\int_0^{\eta}\eta\,\frac{df}{d\eta}\, d\eta$$

Thus,

$$\overline{uv} = 2\delta\overline{U}_M\overline{U}'_M \int_0^{\eta} f^2\, d\eta - 2\delta'\overline{U}_M^2\int_0^{\eta}\eta f\,\frac{df}{d\eta}\, d\eta$$

$$-\delta\overline{U}_M\overline{U}'_M f\int_0^{\eta} f\, d\eta + \delta'\overline{U}_M^2 f\int_0^{\eta}\eta\,\frac{df}{d\eta}\, d\eta$$

$$= \delta\overline{U}_M\overline{U}'_M\left(2\int_0^{\eta} f^2\, d\eta - f\int_0^{\eta} f\, d\eta\right)$$

$$+ \delta'\overline{U}_M^2\left(f\int_0^{\eta}\eta\,\frac{df}{d\eta}\, d\eta - 2\int_0^{\eta}\eta f\,\frac{df}{d\eta}\, d\eta\right)$$

Now, assuming that

$$-\overline{uv} = \overline{U}_M^2\, g(\eta)$$

we get

$$g(\eta) = \frac{\delta\overline{U}'_M}{U_M}\left(2\int_0^{\eta} f^2\, d\eta - f\int_0^{\eta} f\, d\eta\right)$$

$$+ \delta'\left(f\int_0^{\eta}\eta\,\frac{df}{d\eta}\, d\eta - 2\int_0^{\eta}\eta f\,\frac{df}{d\eta}\, d\eta\right) \qquad (4.41)$$

For similarity it must be that

$$\frac{\delta U'_M}{U_M} = C_1 \qquad \text{and} \qquad \delta' = C_2$$

So we obtain only $\delta \approx x$. Note that the gross momentum at any location x is constant, because no forces are involved (no obstacle, no streamwise pressure gradient, etc.), so

$$C_3 = \int_{-\infty}^{+\infty}\rho\overline{U}\overline{U}dy = \rho\int_{-\infty}^{+\infty}\overline{U}_M f\overline{U}_M f\,\delta d\eta = \rho\delta\overline{U}_M^2\int_{-\infty}^{+\infty} f^2\, d\eta$$

$$C_4 = \delta\overline{U}_M^2\int_{-\infty}^{+\infty} f^2\, d\eta$$

$$C_5 = \delta\overline{U}_M^2$$

$$\overline{U}_M \approx \delta^{-1/2}$$

Hence,

$$\overline{U}_M \approx x^{-1/2} \qquad (4.42)$$

If $\delta(x)$ and $U_M(x)$ are known (i.e., the spreading rate), we can find the entrainment (the amount of nonturbulent fluid that becomes turbulent). If $\dot{M} = \dot{M}(x)$ is the amount of fluid entrained by the jet between 0 and x per unit time, then by mass conservation

$$\dot{M}(x) + 2\rho\, d \int_0^\infty \overline{U}(0, y)\, dy = 2\rho\, d \int_0^\infty \overline{U}(x, y)\, dy$$

$$\dot{M}(x) = 2\rho\, d \int_0^\infty (\overline{U}(x, y) - \overline{U}(0, y))\, dy$$

where d is the depth of the jet and ρ is the density of the fluid; alternatively we can write

$$\dot{M}(x) = 2\rho\, d \int_0^\infty \overline{U}(x, y)\, dy - 2\rho\, dh U_0$$

$$U_0 = \frac{1}{h} \int_0^\infty \overline{U}(0, y)\, dy = \text{constant}$$

Now,

$$\int_0^y \overline{U}(x, y)\, dy = \int_0^\eta \overline{U}_M f\, \delta d\eta = \delta \overline{U}_M \int_0^\infty f\, \delta\, d\eta \qquad (4.43)$$

Thus,

$$\frac{\dot{M}(x)}{2\rho\, dh\, U_0} = \frac{\delta(x)\overline{U}_M(x)}{h U_0} \left(\int_0^\infty f\, d\eta - 1 \right) \qquad (4.44)$$

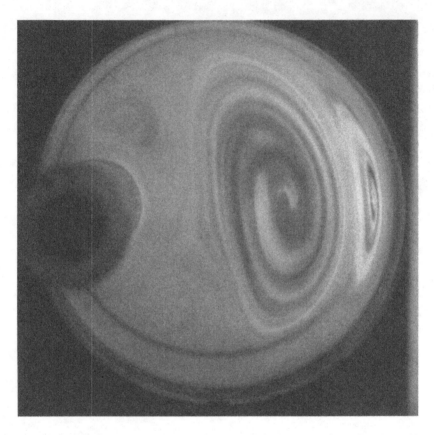

Chaotic mixing between two eccentric cylinders. Photograph from J. Chaiken, R. Chevray, M. Tabor, and Q. M. Tan, *Proc. Roy. Soc. Lond.* A 408, 165–73, 1986.

5 Dynamical Systems and Chaotic Advection

For fluid dynamicists, the word *turbulence* has a precise, albeit not simple, meaning. Turbulence presents disorder that cannot be reproduced in detail, it performs efficient mixing and transport, and it creates vorticity that is irregularly distributed in three dimensions. All these attributes of turbulence can be viewed as the result of random vorticity stretching. That this pseudo-definition excludes wave motion and two-dimensional flows is easily seen from the vorticity equation,

$$\frac{D\omega}{Dt} = (\omega \cdot \nabla)v + \nu \nabla^2 \omega \tag{5.1}$$

which without the first term on the right (null for two-dimensional flows) describes the evolution of a conservative (and diffusible) scalar.

Chaos is an old word, which in its Greek etymology means *emptiness* but which has come to mean *disorder* or *confusion*. In the study of dynamical systems, which is a field to itself and a powerful tool for dealing with the sciences (somewhat akin to but different from mathematics), chaos has yet another meaning. Whenever the motion of a system is fully deterministic in the mathematical sense but is unpredictable in the conventional experimental sense, the motion is termed *chaotic*. This notion of chaos is thus weakly defined; subsequently, we describe it and present criteria for quantifying its sensitive dependence on initial conditions, which is the main attribute of chaos.

Mathematically, in studying chaos we are trying to solve a problem that is represented by the full Navier–Stokes equations (infinite dimensional, nonlinear partial differential equations in the space–time domain) by means of a simpler finite system of ordinary differential or difference equations that describe the conditions in which chaos (being defined only in the time domain) takes place. The idea that intricate systems such as those represented by the Navier–Stokes equations might be amenable to solution of sets of a few simple deterministic ordinary differential equations or difference equations has been entertained only recently, based on the observation that such simple equations exhibit dependence on initial conditions and lead to extraordinarily complex behavior.

121

The field of nonlinear dynamical systems can be divided conveniently into two broad categories: one represented by nondissipative systems (also called hamiltonian systems) and the other closer to turbulence problems, which are dissipative in nature.

5.2 DYNAMICAL SYSTEMS

5.2.a *Hamiltonian systems*

In a nondissipative system, if H is the energy of a particle moving in a potential V, Hamilton's canonical equations for a system of particles are equivalent to Newton's equations and can be written as

$$\dot{q}_k = \frac{\partial H}{\partial p_k} \tag{5.2}$$

$$\dot{p}_k = - \frac{\partial H}{\partial q_k} \tag{5.3}$$

where

$$H(p, q) = \underbrace{\frac{1}{2} |p^2|}_{\text{with multiplicative constants}} + v(q) \tag{5.4}$$

All the q are position vectors and the p are velocity vectors (or momenta). This is classical mechanics. If each p and q has n dimensions, we say that the system has n degrees of freedom (sometimes, but not in this context, reference may be made to such systems as having $2n$ degrees of freedom). Of central importance is the fact that much can be learned about such systems by studying their evolution through a function of F in the $2n$ dimensional space (frequently called *phase space*) that represents the state of the system

$$F = (p_1, \ldots, p_n, q_1, \ldots, q_n) \tag{5.5}$$

It can be shown that a system that is integrable cannot be chaotic. What do we mean by integrability? We say that a Hamiltonian H is integrable if n functions F_i can be found such that $[F_i, H] = 0$; the square bracket is the conventional Poisson bracket, defined as

$$[F_i, H] = \sum_j \left(\frac{\partial F_i}{\partial q_j} \frac{\partial H}{\partial p_j} - \frac{\partial F_i}{\partial p_j} \frac{\partial H}{\partial q_j} \right) \tag{5.6}$$

Another way of defining integrability is by using action-angle variables, but we will not dwell on this here.

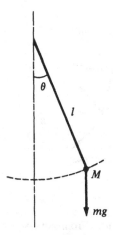

Fig. 5.1 Simple pendulum, schematic. From R. Chevray, "Chaos and the onset of turbulence." In *Advances in Turbulence,* W. K. George and R. Arndt (Eds.). Hemisphere, 1989, New York, pp. 127–158.

Note that although integrable systems cannot be chaotic, nonintegrable systems may or may not be chaotic. One of the earliest and best-known examples of a chaotic nonintegrable system is the three-body problem tackled by Poincaré toward the end of the nineteenth century. It was quite a revelation at the time because it was thought then that most systems in classical mechanics could be integrated. Whether the motion in phase space is regular or irregular depends sensitively on the initial condition.

A single example of such nondissipative systems is the motion of the simple pendulum without friction. With the variables as described in Fig. 5.1, the equation of motion is

$$ml\ddot{\theta} = -mg \sin \theta \qquad (5.7)$$

$$\ddot{\theta} + \frac{g}{l} \sin \theta = 0 \qquad (5.8)$$

which is a second-order nonlinear differential equation; we can then write, with $\dot{\theta} \neq 0$,

$$\left(\ddot{\theta} + \frac{g}{l} \sin \theta\right)\dot{\theta} = 0 \qquad (5.9)$$

$$\frac{1}{2} \dot{\theta}^2 - \frac{g}{l} \cos \theta = \text{constant} \qquad (5.10)$$

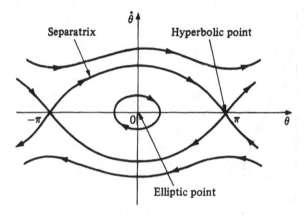

Fig. 5.2 Simple pendulum in phase plane. From same source as Fig. 5.1.

To solve Eq. 5.10 we can also write

$$\dot{\theta} = \frac{\partial H}{\partial p} \tag{5.11}$$

$$\dot{p} = -\frac{\partial H}{\partial \theta} \quad \text{with} \quad p = l\dot{\theta} \tag{5.12}$$

and the corresponding phase-space representation is shown in Fig. 5.2.

$$H(p, \theta) = \frac{1p^2}{2l^2} + \frac{g}{l}(-\cos \theta) = \text{constant} \tag{5.13}$$

$$H(p, \theta) = \frac{1p^2}{2l^2} + \frac{g}{l}(1 - \cos \theta) = \text{constant} \tag{5.14}$$

An important property of the conservation of energy is the conservation of area in phase space. It can be shown that $\delta_{q_i} \delta_{p_i}$ = constant. For three dimensions this corresponds to conservation of volume, and similarly for higher dimensions, because each (p, q) pair corresponds to a constant area.

5.2.b *Dissipative system*

Damping can be added to the frictionless motion of the pendulum, resulting in the following equation, in which k is the damping factor (refer to Fig. 5.3) and the angle θ is assumed to be small:

$$\ddot{\theta} + k\dot{\theta} + \omega^2\theta = 0 \tag{5.15}$$

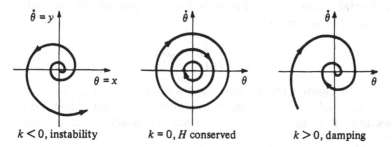

<table>
<tr><td>k < 0, instability</td><td>k = 0, H conserved</td><td>k > 0, damping</td></tr>
</table>

Fig. 5.3 Phase portrait of Eq. 5.15. From same source as Fig. 5.1.

where

$$\omega^2 = \frac{g}{l} \tag{5.16}$$

Energy: $H = \frac{1}{2}\dot{\theta}^2 + (g/l)(1 - \cos\theta)$

For small θ,

$$H = \frac{1}{2}\dot{\theta}^2 + \frac{g}{l}\frac{\theta^2}{2} = \frac{1}{2}\left(\dot{\theta}^2 + \frac{g}{l}\theta^2\right) = \frac{1}{2}(\dot{\theta}^2 + \omega^2\theta^2)$$

$$\frac{dH}{dt} = (\ddot{\theta} + \omega^2\theta)\dot{\theta} = -k\dot{\theta}^2 \tag{5.17}$$

H is conserved only when $k = 0$. When $k > 0$, $\dfrac{dH}{dt} < 0 \rightarrow$ damping;

and when $k < 0$, $\dfrac{dH}{dt} > 0 \rightarrow$ unstable. The phase portrait for small θ is given by

$$y = \dot{\theta} \tag{5.18}$$

$$\dot{y} = \ddot{\theta} = -k\dot{\theta} - \omega^2\theta \tag{5.19}$$

$$x = \theta \qquad \dot{x} = y \tag{5.20}$$

$$\dot{y} = -ky - \omega^2 x \tag{5.21}$$

In dissipative systems, we have contraction of area in phase space.

van der Pol equation: If we now vary k so that it is negative for small amplitudes but positive for large ones, then

$$k = -k_0\left(1 - \frac{\dot{\theta}^2}{\theta_0^2}\right) \tag{5.22}$$

will satisfy this requirement; then we get a phase portrait corresponding to the van der Pol equation,

$$\ddot{\theta} - k_0\left[1 - \frac{\theta^2}{\theta_0^2}\right]\dot{\theta} + \omega^2\theta = 0 \qquad (5.23)$$

and the corresponding phase-plane trajectories shown in Fig. 5.4. There exists a limit cycle, where we cannot identify the particular initial conditions and where all initial conditions will eventually lead to this limit cycle.

5.3   QUANTIFYING CHAOS

5.3.a   *Logistic map*

This is now a well-known two-dimensional map represented by

$$x_{n+1} = rx_n(1 - x_n) \qquad (5.24)$$

For a value of r above a certain threshold (about 3.6); as an example, we let $r = 4$ (Fig. 5.5): this sequence is known to be chaotic, and the orbits diverge exponentially fast on the average. The bisector is used in the figure as a means of constructing the next point in the sequence.

5.3.b   *Lyapunov exponent*

If we start with x_n and with a point very close to it, $x_n + \delta x_n$, the next point will be

$$x_{n+1} + dx_{n+1} = f(x_n + dx_n)$$

$$\approx f(x_n) + dx_n \cdot \frac{\partial f(x_n)}{\partial x}$$

so $dx_{n+1} = dx_n\,(\partial f/\partial x)\,(x_n)$. Therefore,

$$dx_{n+1} = dx_0 \prod_{i=0}^{n} \frac{\partial f}{\partial x}\,x_i$$

For the logistic map,

$$dx_{n+1} = dx_0 \prod_{i=0}^{n} r[1 - 2x_i]$$

For n large enough, if $|r(1 - 2x_i)| > 1$, the logistic equation is chaotic for all r. Thus, we can define a Lyapunov exponent λ to make dx_n grow

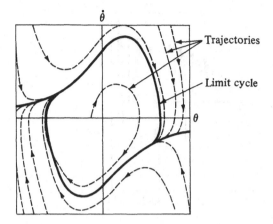

Fig. 5.4 Phase plane of van der Pol equation. From same source as Fig. 5.1.

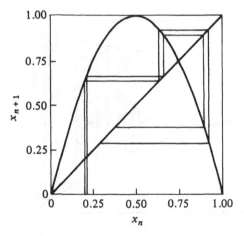

Fig. 5.5 Logistic map. From same source as Fig. 5.1.

exponentionally fast: $dx_n = dx_0\,2^{\lambda n}$, for this we need

$$\lambda = \lim_{n \to \infty} \frac{1}{n} \sum_{i=0}^{n-1} \log_2 |f'x_i| \qquad (5.25)$$

In the bifurcation regime in which orbits branch out, $3 < r < r_\infty$, $\lambda = 0$ only at bifurcation points, as shown in Fig. 5.6; and in the chaotic regime $r_\infty < r < 4$, some windows of $\lambda < 0$ exist, as shown in Fig. 5.7. In the bifurcation regime $r > 3$, we first have one fixed point (1 cycle), then above $r \simeq 3.499$, we have 2 cycles; and as r increases, we go to 4

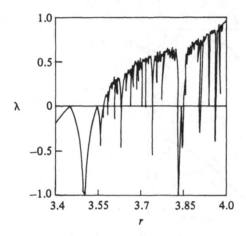

Fig. 5.6   Logistic map, variation of the Lyapunov exponent with *r*. From same source as Fig. 5.1.

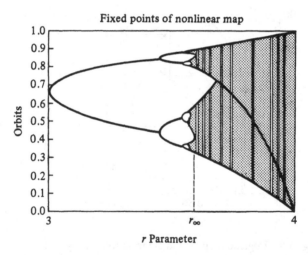

Fig. 5.7   Iterates of the logistic map. Same source as Fig. 5.1.

cycles, and so on. All these cycles represent pitchfork bifurcations. A special case for $r \simeq 3.1$ is shown in Fig. 5.8.

5.3.c   *Fractals: How long is the coast of Portugal?*

The length of a coastline is a function of the "yardstick"; it increases as the yardstick becomes smaller (see Fig. 5.9). The length of the coast

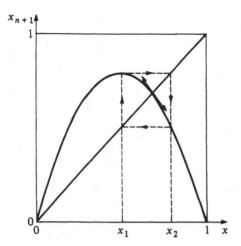

Fig. 5.8 Logistic map, particular case. Same source as Fig. 5.1.

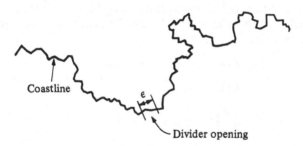

Fig. 5.9 Model of a coastline. Same source as Fig. 5.1.

of Portugal as estimated by Portugal is greater than the length estimated by Spain because a smaller yardstick is used (map is more enlarged); see Fig. 5.10. For instance, Portugal gives the length of her coastline as 1214 km whereas Spain gives Portugal's coastline length as 987 km.

To measure the length L on a map, we set our divider opening at ε. Then

$$L(\varepsilon) = \text{number of steps} \times \varepsilon$$

As ε decreases, we might expect $L(\varepsilon)$ to approach a value corresponding to the true length. In fact, this is not the case. Instead, $L(\varepsilon)$ increases without limit, and the reason is obvious: To approximate the length of a coastline we need to have $f\varepsilon^{-D}$ intervals of length ε (where f and D are arbitrary constants) so that if we have a straight line then $D = 1$,

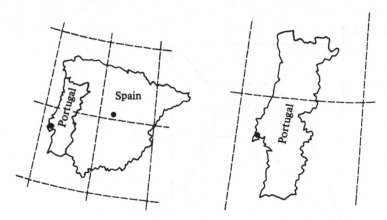

Fig. 5.10   Portugal within Spain. Same source as Fig. 5.1.

and otherwise D depends on the physical nature of the coastline. However, it turns out that D has a universal value for coastlines!

The snowflake shown in Fig. 5.11 can be taken as a model for a
coastline. In this case,

$$L(\varepsilon) = l\varepsilon^{-D} \times \varepsilon = l\varepsilon^{1-D}$$

$$L(1) = l$$

$$L\left(\frac{\varepsilon}{3}\right) = \frac{4}{3}L(\varepsilon) \qquad\qquad (5.26)$$

because, when we change the dividers from ε to $\varepsilon/3$, we obtain

$$\varepsilon \rightarrow L(\varepsilon)$$

$$\frac{\varepsilon}{3} \rightarrow \frac{4}{3}L(\varepsilon)$$

but

$$L\left(\frac{\varepsilon}{3}\right) = l\,\varepsilon^{1-D}3^{D-1}$$

$$3^{D-1} = \frac{4}{3} \rightarrow 3^D = 4$$

$$D\log 3 = \log 4 \rightarrow D = \frac{\log 4}{\log 3} \qquad\qquad (5.27)$$

$$D \approx 1.26$$

Fig. 5.11   Koch curve (snowflake), constructed by dividing each segment into three equal parts and building an equilateral triangle on the central segment, and then repeating this procedure. From same source as Fig. 5.1.

Thus, looking again at the coast of Portugal, we see that

$$\frac{L(\text{as measured by Spain})}{L(\text{as measured by Portugal})} = \frac{L(\varepsilon_1)}{L(\varepsilon_2)}$$

where $\varepsilon_1 > \varepsilon_2$. Then, $\varepsilon_2 = \varepsilon_1/\sqrt{A_1/A_2} = \varepsilon_1/2.34$, and $L(\varepsilon_1) = L(\varepsilon_2) \cdot (0.80)$ because $(2.34)^{-0.26} = 0.8$, and from various encyclopedias we have $L(\text{Spain})/L(\text{Portugal}) = (987 \text{ km})/(1214 \text{ km}) = 0.81$. The two values are indeed very close.

Both Spaniards and Portugese work on maps of the same physical size and probably use the same method (same dividers) to determine the length. They consequently arrive at different values! In general, we say that a mathematical entity that has a fractal (Hausdorff–Besicovitch) dimension larger than its topological dimensions is a *fractal*, a term coined by Mandelbrot from the latin *frangere* "to break." We must be careful about *fractal dimension*: The term does not imply that the exponent is fractional. In fact, the projection of Brownian motion is a fractal with dimension 2 because it is space-filling in the plane, but its topological dimension is obviously only 1. Formally, we can propose the following definition for the Hausdorff–Besicovitch dimension: If, for a set of points in d dimensions, the number $N(l)$ of d-spheres of diameter l needed to cover the set increases with $N(l) \propto l^{-D}$ for $l \to 0$, then D is the Hausdorff dimension. As another example, a straight line is obviously not a fractal, but a turbulent signal given as a function of time is a fractal.

5.4   STRANGE ATTRACTORS AND THE LORENZ SYSTEM

In chaotic systems, the orbits (trajectories of points having the position and velocity as coordinates) lie on a geometrical surface called an *attractor*. The simplest attractor (to which orbits converge) is the fixed point representing a system at rest. A periodic regime is represented by a limit cycle, and all periodic systems with more than one period can be described by *tori*. Whenever the dimension of the attractor is fractal, it is termed a *strange attractor*. The notion of attractor is indissolubly linked to that of dissipation. It is also easy to see that dissipation leads to contraction of areas (or volumes) in phase space and that the dimension of the attractor is always lower than that of the phase space. We must satisfy two apparently contradictory conditions in the phase space: convergence toward the attractor and divergence (exponential) on the attractor itself. Moreover, there exists a topological constraint: The phase space is bounded, and trajectories cannot intersect. In many instances, instead of studying the trajectories in phase space, we investigate their intersection by a plane. Such a representation has one less dimension than the original phase space and is referred to as a *Poincaré section* (or surface of section). Lorenz, in his attempts to study numerically convective motions in the atmosphere, had to solve a system of three nonlinear partial differential equations describing the Rayleigh–Benard problem of fluid heated from below in a rectangular box. By severely truncating the solution expanded in Fourier modes and keeping the dependent variables x, y, and z, representing the three lowest modes, he reduced this system to that of the following three coupled, nonlinear, first-order differential equations:

$$\dot{x} = \sigma(y - x)$$
$$\dot{y} = x(r - z) - y$$
$$\dot{z} = xy - bz \qquad\qquad (5.28)$$

The parameters σ, b, and r correspond to the Prandtl number, a geometric factor, and the Rayleigh number, respectively. Among the many solutions provided as the parameters are varied, Lorenz chose to study those corresponding to $\sigma = 10$ and $b = \frac{8}{3}$ as r takes on positive values. The solution corresponding to this particular system is shown on the phase portrait of Fig. 5.12.

The structure of this attractor is highly layered, and the dimension of the attractor is 2.06. All trajectories spiral around one of the two attracting points for some time and then switch to the other in a seemingly haphazard yet deterministic manner. The fact that two trajectories that are initially very close will separate exponentially, on the average, is one of the attributes of deterministic chaos. This chaotic behavior is described in detail by Lorenz himself, who commented that such behavior has philosophical and practical consequences for long-range weather forecasting. The difficulty of making long-term weather forecasts is a good illustration of this remark.

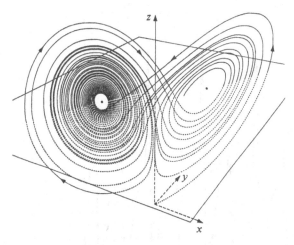

Fig. 5.12   Lorenz attractor. From same source as Fig. 5.1.

5.5   CHAOS IN FLUIDS

5.5.a   *Chaotic advection*

An important concept in the description of fluid motion is that of *chaotic advection* (a term coined by Aref), which is also called lagrangian turbulence. A simple example is that of the velocity in a two-dimensional shear layer (Fig. 5.13).

We note in passing that our method of describing a flow influences our understanding of it. For instance, in the lagrangian description (Fig. 5.14), in which every particle of a flow is identified and tracked, we perceive the flow in a different manner than we do in an eulerian frame of reference, in which we observe the particles passing by a fixed point. The eulerian method of description has the effect of scrambling the information.

Although the measured longitudinal component of the velocity at a given point in the laminar region of a flow from a hot-wire signal looks periodic (see Fig. 5.13), such is not the case for the corresponding particle velocity in the lagrangian description (Fig. 5.14). It is observed, actually, that the centers of vortices rotate about each other in a periodic manner. That such a flow, with a periodic stream function (hamiltonian as we shall see), generates chaotic particle trajectories is rather curious; however, such an observation is very much akin to the corresponding observations of numbered-ball trajectories in the steadily rotating drums of some lottery games.

A simple example, which (like the previous one), has been studied in the laboratory and investigated theoretically, is the two-dimensional

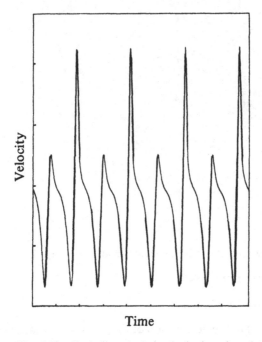

Fig. 5.13   Periodic eulerian velocity in a shaer layer. From same source as Fig. 5.1.

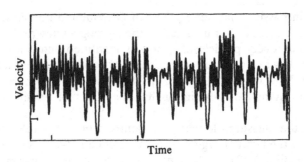

Fig. 5.14   Lagrangian velocity in a shear layer. From same source as Fig. 5.1.

flow in a horizontal plane between two vertical eccentric cylinders that rotate alternately. The gap between the cylinders is filled with a glycerine solution, and the corresponding Reynolds number (based on the average gap width and the circumferential speed of the inner cylinder) is approximately 0.1, well within the Stokes flow regime. The particle trajectories are governed by the following system of ordinary differential equations,

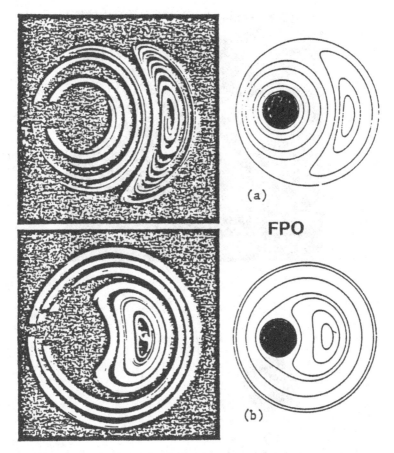

(a)

FPO

(b)

Fig. 5.15 Streamlines obtained experimentally for (a) inner cylinder rotating and (b) outer cylinder rotating. From J. Chaiken, R. Chevray, M. Tabor, and Q. M. Tan, *Proc. Roy. Soc. Lond.* A 408, 1986, 165–73.

$$\dot{x} = u(x,\, y,\, z,\, t)$$

$$\dot{y} = v(x,\, y,\, z,\, t) \tag{5.29}$$

which form a nonautonomous, hamiltonian system with one degree of freedom. Note that the stream function ψ satisfying

$$\dot{x} = -\frac{\partial \psi}{\partial y} \qquad \dot{y} = -\frac{\partial \psi}{\partial x} \tag{5.30}$$

acts as a hamiltonian for which the real-space coordinates x and y are the canonically conjugate phase-space variables. Whenever this system is time independent, then, as we have seen, it is integrable and therefore

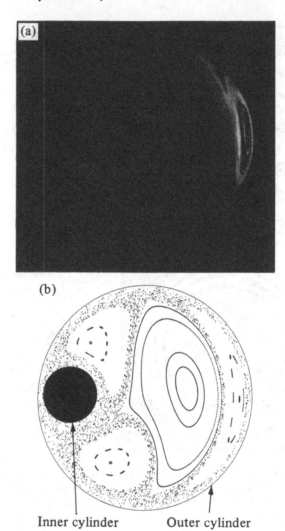

Fig. 5.16   Poincaré section corresponding to the basic cycle consisting of anticlockwise rotation of the inner cylinder followed by one complete clockwise rotation of the outer cylinder: (a) experimental results and (b) numerical results. From same source as Fig. 5.1.

cannot be chaotic. For nonautonomous systems (which contain time explicitly), such as the one we have just described, the time dependence comes into play. The introduction of discontinuities in the motion of the cylinders, which are alternatively rotated through a set angle in either direction, causes fluid particles to follow trajectories that alternate from one system of streamlines (Fig. 5.15a) when the inner cylinder rotates

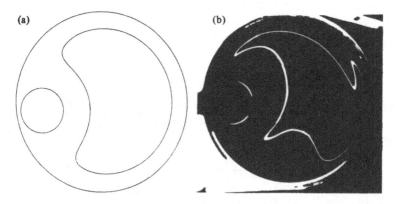

Fig. 5.17 (a) Initial contour generated by rotation of the outer cylinder and (b) experimental result after 15 periods. From same source as Fig. 5.15.

to the other (Fig. 5.15b) when the outer cylinder rotates. For each rotation, the flow is steady and pathlines coincide with streamlines and streaklines; but as soon as the particles switch from one system to the other, chaotic motion can ensue.

In Figs. 5.16a and 5.16b, we show the successive locations (Poincaré section) of eleven particles at the end of each of 800 cycles. The dyed particles photograph (Fig. 5.16a) was taken after only 349 cycles. Clearly discernable in this figure is that in regions where the streamline patterns are markedly different for the two steady cases, we expect the orbits (trajectories) to be chaotic. In two regions near the inner cylinder, the particle trajectories are very regular; these are called *elliptic regions* because the flow curls around elliptic fixed points. The large area on the right is obviously a KAM curve, named after Kolmogorov, Arnold, and Moser, who proved the KAM theorem, which states that the motion in the phase space of classical mechanics is neither completely regular nor completely irregular, but that the type of trajectory depends sensitively on the initial conditions chosen. In the other places the dye has spread very uniformly; these are the chaotic regions. Such regions are of much interest for engineering applications because it is there that we can get very significant mixing even in very low Reynolds number flows. To indicate how an initial streamline (Fig. 5.17a) is deformed, we show the resulting pattern obtained after only 15 periods (Fig. 5.17b).

We now present a numerical study of two rows of vortices of different circulation in an inviscid fluid. The complex potential corresponding to this flow, which is stable except when the separation distance between

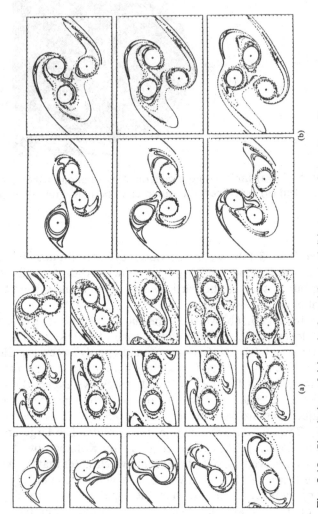

Fig. 5.18   Simulation of (a) a single pairing and (b) a successive vortex. From same source as Fig. 5.1.

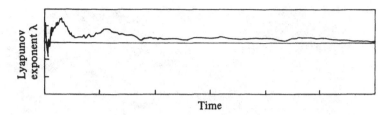

Fig. 5.19 Lyapunov exponent for chaotic advection in the two-dimensional shear layer. From same source as Fig. 5.1.

vortices is perturbed, can be written as

$$f(z) = f_1(z) + f_2(z) = \frac{i\Gamma_1}{2\pi} \ln \left[\sin \frac{\pi}{2\,1} (z - z_1) \right]$$

$$+ \frac{i\Gamma_2}{2\pi} \ln \left[\sin \frac{\pi}{2\,1} (z - z_2) \right] \qquad (5.31)$$

and, after a perturbation is introduced, it can be written in the form describing the trajectories of the particles:

$$\dot{X} = -\frac{\partial \Psi}{\partial Y} (X, Y, T)$$

$$\dot{Y} = \frac{\partial \Psi}{\partial X} (X, Y, T) \qquad (5.32)$$

This corresponds to a two-dimensional hamiltonian nonautonomous system that may be chaotic. Note that this evolution is nondissipative, the area is conserved in the phase plane, and consequently there is no attractor. Whereas in a viscous flow the evolution of the velocity field is governed by partial differential equations (the Navier–Stokes equations), in an inviscid flow the particle trajectories are obtained from a set of coupled ordinary differential equations. Numerical integration of this system as time varies is shown graphically in Fig. 5.18a for a single pairing and in Fig. 5.18b for successive pairings. Note in these figures the spread of particles initially set on a straight line parallel to the axis of the row of vortices; their motion is clearly chaotic even though the centers of vortices turn periodically around each other. The Lyapunov exponent for this system has been computed to be equal to .31 and is presented in Fig. 5.19, confirming the chaotic nature of the flow.

The development of grid generated turbulence at moderate Reynolds number. Reproduced by permission from "An Album of Fluid Motion."

6 Chaos and the Onset of Turbulence

We have given in Chapter 5 a format for describing turbulence and also a pseudo-definition of chaos. The main attribute of chaotic systems is that they can be described by a system of nonlinear differential equations. The field of nonlinear dynamical systems can be divided conveniently into two broad categories: those represented by nondissipative systems and those closer to the real turbulence problems, which are dissipative. Now we briefly touch on the nondissipative systems, which lead to interesting studies of spatial complexity, spend a little more time on dissipative systems, insofar as they lead to the onset of turbulence, and speculate on their value for trying to describe fully developed turbulent states.

As we saw, hamiltonian systems have phase portraits with non-bounded orbits because these trajectories are not brought together by an attractor. For such systems, the energy is constant on each trajectory; it is an integral of the system. We also saw a well-known example of such a system, the simple pendulum with no friction.

In general, if we consider a dynamical system described by an equation such as

$$\frac{d\mathbf{x}}{dt} = \mathbf{F}(\mathbf{x}) \tag{6.1}$$

or

$$\mathbf{x}_{k+1} = \mathbf{F}(\mathbf{x}_k) \tag{6.2}$$

corresponding, respectively, to continuous differential equations and to difference equations in the vector space with N dimensions to which $\mathbf{x}$ and $\mathbf{x}_k$ belong.

If we assume that $\mathbf{x}$ at time t is a continuous function of an initial value $\mathbf{x}_0$, then for an error $\delta\mathbf{x}_0$ on the position x_0, we can get δx at time t as small as we want by choosing a correspondingly small $\delta\mathbf{x}_0$. In other words, for $\delta\mathbf{x}_0$ bounded, we can get $\delta\mathbf{x}_t$ growing without bound with time. This operation translates into mathematical terms what is usually called *sensitivity to initial conditions*. The dissipative systems studied exclude cases for which this sensitivity to initial conditions is limited either to the choice of $\mathbf{x}_0$ or to the case for which $\mathbf{x}(t)$ goes to infinity.

141

Furthermore, we restrict ourselves to the study of systems for which the growth rate of the error δx_t is exponential: that is, chaotic systems.

6.2   ROUTES TO TURBULENCE

The phenomenon first observed by Bénard and interpreted by Rayleigh is that of the flow between two parallel horizontal plates (Fig. 6.1), caused by heating the lower plate at a temperature $T_0 + \Delta T$ above the temperature T_0 to which the lower plate is held. No flow is observed for low values of ΔT; the transfer of heat necessary to maintain the temperature difference is achieved solely by conduction.

For values ΔT_1 or ΔT greater than a critical value ΔT_c, the necessary transfer cannot be achieved by conduction alone. Convective rolls set in that greatly increase the transfer of heat and permit the system to reach an equilibrium state. The rolls in that state are of about the same diameter as the fluid depth and are obviously counterrotating. Such a transition from the conductive state to the convective counterpart is termed a *bifurcation*. The concept of bifurcation is difficult, but in a loose sense, whenever the solution of a system of equations (for a given value of a parameter) changes qualitatively (as depicted for instance on the phase space), we call such a location in space a *bifurcation point* (whether it leads to one or several branches) and the corresponding value of the parameter is the *critical value*. Whenever, the solution leads to a limit cycle after going through the critical value of the parameter, the bifurcation is called a *Hopf bifurcation*. At a still greater temperature difference, ΔT_2, a new instability sets in, this time in the third dimension, which allows yet a greater heat transfer. This mode and the preceding mode are clearly visible on the spectrum shown in Fig. 6.2. This stage corresponds to the second bifurcation. A third bifurcation is observed beyond this point for still greater temperature difference, $\Delta T_3 > \Delta T_2$, and turbulence as we know it is observed throughout the flow. As can be surmised, the temperature difference alone is a unique criterion characterizing each successive stage. Other parameters that characterize the geometry and the physical properties of the fluid combine to form the Rayleigh number

$$Ra = \frac{\rho^2 C_p g \beta d^3 \Delta T}{\mu k}$$

in which ρ, β, C_p, μ, and k are physical characteristics of the fluid, d is the width of the plates, and g the gravitational attraction. This flow lacks the time dependence in the initial stage, which distinguishes it clearly from other hydrodynamic instabilities that lead directly to a turbulent state with a statistically fluctuating time dependence.

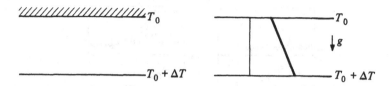

Fig. 6.1   System at rest: heat transfer due to convection alone

As the Rayleigh number Ra varies, we first see (Fig. 6.2a) one frequency f_2 appearing together with its harmonics; then (Fig. 6.2b) a second frequency f_1 is visible together with the combinations $(k_1 f_1 \pm k_2 f_2)$. Finally, as chaos sets in (Fig. 6.2c), sharp lines are clearly evident; these then disappear to give the broadband spectrum of Fig. 6.2d.

Another well-known experiment is that of Taylor flow, first described by Couette and then analyzed in detail by Taylor. In this experiment, the gap between two concentric cylinders is filled with fluid and the inner cylinder is rotated. Because the outer cylinder is stationary, we first observe the entrainment of the fluid between the cylinders. For low values of the velocity (low Reynolds numbers, Re), we obtain a laminar, well-ordered flow that can be calculated by solving the Navier–Stokes equations to yield

$$V_\theta = Ar + \frac{B}{r}$$

in which A and B are determined by the boundary conditions and there is no radial component of the velocity. As the Reynolds number increases, the next phase is called time-independent Taylor vortex flow, which consists of toroidal vortices superimposed onto the flow. In 1923 G. I. Taylor observed the formation of these vortices above a certain Reynolds number. He performed a linear stability analysis to determine how the effects of viscosity delayed this phenomenon. As the Reynolds number increases from this point, time-dependent wavy vortex flow is observed. This flow consists of Taylor vortices superimposed on traveling waves in the azimuthal direction.

Experimental results for this flow are presented in Fig. 6.3. For a Reynolds number equal to the critical number (119), Taylor vortices occur. This flow is time independent and no frequencies are shown. A further increase in velocity up to $Re/Re_c = 1.2$ gives a single sharp frequency in the power spectra; this frequency ω_1 corresponds physically to the azimuthal waves described earlier. As the Reynolds number increases to $Re/Re_c = 10.06$, a second frequency ω_3 is observed. The ratio of the two frequencies ω_3/ω_1 increases continuously and monotonically as the Reynolds number increases. This ratio is irrational, and the flow

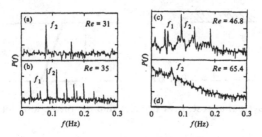

Fig. 6.2　Power spectra in the Rayleigh–Benard experiment. Adapted from H. L. Swinney and J. B. Gollup, *Phys. Today* **8**, 41, 1978.

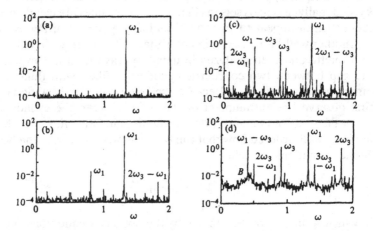

Fig. 6.3　Velocity fluctuation power spectra for Taylor vortices. Adapted from P. Fenstermacher, H. Swinney, and J. Gollub, *J. Fluid Mech.* **94**, 103, 1979.

is described as quasi-periodic. At $Re/Re_c = 12$, the first chaotic element appears; it corresponds to a frequency B, and it suffers from hysteresis. This frequency is observed to continue while the noise level appears to be rising up to its level on the power spectra. The two frequencies ω_3 and ω_1, however, disappear at $Re/Re_c = 19.3$ and 21.9, respectively.

6.3　ROUTES TO CHAOS

Perhaps the best-known route to chaos is that postulated by Ruelle, Takens, and Newhouse (1978). Whereas Landau (1944) had suggested that chaos appears after a series of incommensurable frequencies generated by an infinite sequence of Hopf bifurcations, leading to a discrete spectrum that becomes continuous only after an infinite sequence of bifurcations, Ruelle and Takens (1971) proposed a totally different approach. Together with Newhouse (1978) they postulated that regardless of the phase-space dimension, the sequence of bifurcations toward chaos as the control parameter is varied is as follows: fixed points, limit cycle, tori (2 and 3 torus), and, finally, strange attractors. After a finite number of bifurcations (sometimes two, but usually three) corresponding to a 2 or 3 torus, the system becomes unstable and develops a strange attractor. Chenciner and Ioos (1979) pointed out that the transition from a 2 torus to a 3 torus is a rare event, requiring very special circumstances. Newhouse, Ruelle, and Takens have shown that a 3 torus can be perturbed in such a way as to give rise to a "stable" chaotic motion.

Note that this route is not generic of the transition phenomenon but is a purely mathematical development that coincides with experimental observations such as those of the Rayleigh–Bénard problem.

A scenario recently suggested by M. Feigenbaum (1978) is called the period-doubling route to chaos. This is the route described by the logistic map we presented in Chapter 5. When the slope of a function is strictly greater than one at its point of intersection with the bisectrix, the sequence $x_0, f(x_0), f^2(x_0), \ldots, f^m(x_0), \ldots$ does not lead to a single value but to two distinct values. As the parameter r increases, the orbit becomes unstable, and a new quasi-periodic orbit of half-frequency appears. Such a map (Fig. 6.4) shows, as r increases, the period-doubling route to chaos. For still higher values of r, windows of order within the chaotic regions exist. The Rayleigh–Bénard problem is known to exhibit such a route for a certain range of Rayleigh numbers.

The last known route to chaos, described by Manneville and Pomeau (1979), is characterized by intermittent bursts of chaos arising in an otherwise regular signal. Intermittency associated with low-order dynamical systems occurs when trajectories continually visit unstable points in a bounded phase space; additionally, these unstable points must be, in some sense, near their stability threshold. At first, the trajectory remains near the unstable fixed point (relative quietness), and then it is thrown out into the whole of the phase space; the result is chaos. How long the trajectory remains in its quiet region has been predicted by Manneville and Pomeau to be a function of how close the unstable point is to its stability threshold.

The transition from the periodic motion of the limit cycle to the chaotic motion of the strange attractor is known to occur in the Lorenz system when r is near 166. For values of r below this limit, the trajectories are attracted to a stable limit cycle; for values above 167, a strange attractor exists. Such a transition is depicted in Fig. 6.5. A Poincaré section for the plane $x = 0$ is shown in Fig. 6.6. Because this section appears to approximate a smooth curve, the plot of y_{n+1} versus y_n is shown in Fig.

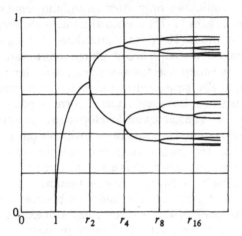

Fig. 6.4   Iterates of the logistic map. From same source as Fig. 5.1.

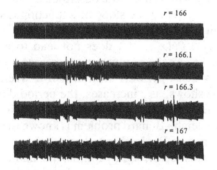

Fig. 6.5   Lorenz model: one coordinate as a function of time. From same source as Fig. 5.1.

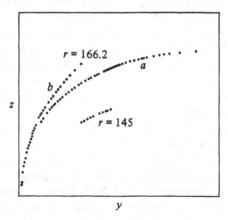

Fig. 6.6   Poincaré section for $x = 0$

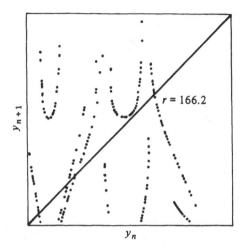

Fig. 6.7   Poincaré map for $r = 166.2$

6.7, an enlargement of which (in the region where the parabolic section crosses the bisectrix) is shown in Fig. 6.8. On this enlargement, the trajectory is tracked as the particle moves through the channel, and a periodic signal results. If the particle leaves the channel, then it can explore much of the phase space, as shown in Fig. 6.7; this corresponds to the chaotic state. Because the phase space is bounded, it is likely that the trajectory will pass through the channel several times. It is clear in Fig. 6.7 that the behavior of the solution is controlled by the relative position of the bisectrix and the curve, the latter being a function of the parameter r.

Just as there are three ways for a point to become unstable according to linear stability theory, Manneville and Pomeau have identified three types of intermittency, each type being characterized by the type of instability of the linearized system. In type I, intermittency occurs when the unstable point is due to one eigenvalue of the linearized system crossing the unit circle on the positive real axis (Fig. 6.9a). In such a case, the laminar signal increases monotonically because of the real and positive values of the eigenvalue. Type I intermittency is observed in the Lorenz system and also in the logistic map. For systems in which the eigenvalues are complex (such as systems with at least two degrees of freedom) and simultaneously cross the unit circle, the resulting signal is oscillatory (Fig. 6.9b). This intermittency (type II) has been observed recently in the Belousov–Zhabotinsky reaction by Roux, Kepper, and Swinney (1984). Finally, when one eigenvalue of the linearized system crosses the unit circle on the negative real axis (Fig. 6.9c), type III intermittency occurs. This type of intermittency has been observed

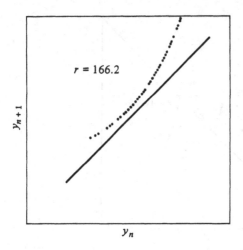

Fig. 6.8   Enlargement of Fig. 6.7

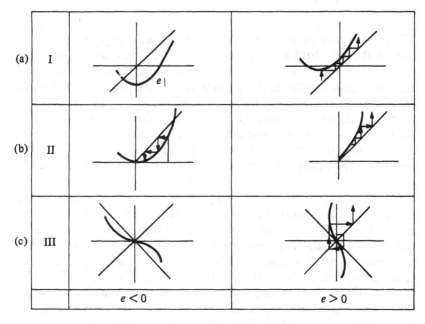

Fig. 6.9   The three types of intermittencies. From same source as
Fig. 5.1.

experimentally for the first time in the Rayleigh–Benard system by
Dubois, Rubio, and Bergé (1983).

Intermittency theory yields one of the few quantitative results predicting the average length of time a "laminar" signal is expected to last

before bursting into chaos. We have seen that in type I intermittency, the Poincaré map y_{n+1} versus y_n is parabolic and is represented locally by

$$y_{n+1} = y_n + e + a(y_n^2) \qquad (6.3)$$

Herein, the parameter e indicates how far the system is disturbed from stability (i.e., the width of the channel). Noting that y is small throughout the channel, Eq. 6.3 can be replaced by the differential equation

$$\frac{dy}{dn} = e + ay^2$$

This leads to a closed-form solution, from which we can show that the number of steps needed to cross the channel is inversely proportional to the square root of e. This result is in excellent agreement with numerical predictions for the Lorenz system as well as for the logistic map. This relationship between the length of the laminar signal and e hold for all systems displaying type I intermittency. Similar studies were carried out for type II and type III intermittencies. Here, as before, we approximate the Poincaré map with a simple difference equation, which is cubic for both types. Again, Mannevile and Pomeau found that the length of time expected for the laminar region is inversely proportional to e for these types II and III.

An excellent example of intermittency can be seen in the game of pinball. A plot of the level of the ball as a function of time would look chaotic, especially if we consider the ultimate fate of two balls that initially have indistinguishable locations and velocities. However, if the ball gets trapped between a wall and a post, a pleasing experience indeed, the ball oscillates in the "channel" until it rolls down. The ball then moves randomly through the field until it lands again (with luck) in a channel. Such seems to be the way intermittency is generated in low-order dynamical systems such as the Lorenz equation, the logistic map, and the pinball.

6.4 CHAOS QUANTIFICATION

It is important to establish criteria that enable us to determine whether a system is chaotic, and also important to establish an order or a degree of "chaosness" among several chaotic systems. When we are analyzing random signals, we often perform a power spectrum or Fourier decomposition of a signal. In practice, our record being finite, we perform a Fourier transform on this sample (more often than not, using fast Fourier transform, the program of which is readily available in statistical packages). A characteristic power spectrum can be associated with each type of signal. For a periodic signal, for instance, we obtain discreet frequencies $f_1 = \dfrac{1}{T}$ and the harmonic $2/T$, $3/T$, ... For a quasi-periodic signal, a number of fundamental frequencies $f_1, f_2, \ldots, f_n$ are necessary to specify it in phase space, and its orbits are evolving on an n torus. In general, therefore, we can observe $f_1, f_2, \ldots, f_n$ as well as harmonics:

$$f = k_1 f_1 + k_2 f_2 + \cdots + k_n f_n$$

Whenever the frequencies are commensurable, we have a periodic signal of period T corresponding to a spectrum consisting of delta functions separated by a frequency at least equal to $1/T$.

If, on another hand, the frequencies are not commensurable, then we have a dense distribution of delta functions. Finally, when the signal is chaotic (which we cannot in general distinguish from a random signal), then the spectrum is continuous. Even the distinction between a chaotic and a quasi-periodic signal with incommensurable frequencies is sometime difficult to make. Figure 6.10 shows such a signal classification.

Another function commonly used in turbulence research which is appropriate for the analysis of chaotic signals is the autocorrelation function, defined as

$$R(\tau) = \lim_{T \to \infty} \frac{1}{T} \int_0^T x(t) x(t + \tau) \, dt$$

This function provides an indication of coherence of the signal over times of order τ. If large-scale structures are embedded in the signal, then the autocorrelation takes on the shape of a damped periodic function. For chaotic systems, as we observe a loss of internal similarity with time, the autocorrelation function drops quickly to zero. Note that no information is provided by the autocorrelation because it is the Fourier–Stiljes integral transform of the spectrum.

To reduce the dimension of the phase space (by one), we make Poincaré sections, which enable us to visualize in a space of lower dimension the form of the attractor on which the orbits lie. In three-dimensional phase space, for instance, a cut by a plane readily shows a Poincaré section. The ensemble of points thus obtained permits us to observe the nature of the orbits. Obviously, a periodic orbit would result in a single point whereas a quasi-periodic orbit with commensurable frequency would correspond to a finite number of points in the plane, always visited in the same order. In the noncommensurable-frequencies quasi-periodic system, the Poincaré section is a dense line corresponding to a plane section of the torus on which the orbits lie. Finally, a chaotic system has a Poincaré map showing a complex structure corresponding to strange (layered) attractor sections, as shown in Fig. 6.11.

One of the more common ways to ascertain whether chaos exists in

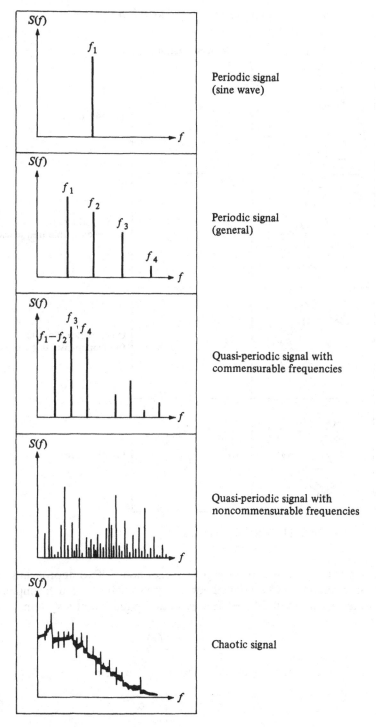

Fig. 6.10 Signal classification

Orbits   Poincare sections

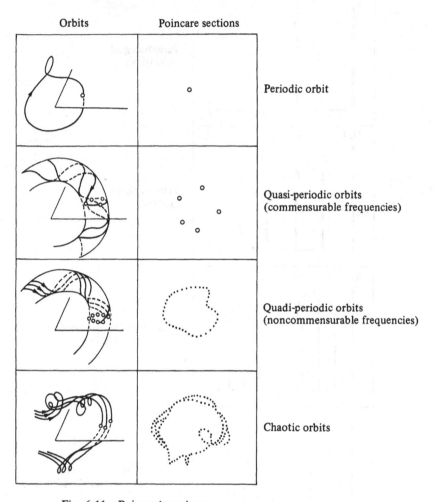

Periodic orbit

Quasi-periodic orbits
(commensurable frequencies)

Quadi-periodic orbits
(noncommensurable frequencies)

Chaotic orbits

Fig. 6.11   Poincaré sections

a system is to check the Lyapunov exponents. Two approaches exist:
matrix analysis and ergodic theory. Matrix analysis is much simpler and
is the one presented here. The evolution equation of a system

$$\frac{dx}{dt} = F(x) \tag{6.4}$$

can be linearized around the solution $x(t)$ to yield

$$\delta\dot{x} = \left.\frac{\partial F}{\partial x}\right|_{x(t)} \delta x \tag{6.5}$$

where $\partial F/\partial x$ is the matrix of the partial derivative corresponding to the system represented by Eq. 6.4. Equation 6.5 can be integrated numerically to find a new matrix $L(t)$ such that

$$\delta x(t) = L(t)\,\delta x(0)$$

in which $\delta x(0)$ and $\delta x(t)$ are the errors at times 0 and t. Thus, the behavior of a small initial error is given by the asymptotic behavior of $L(t)$. It is therefore clear that the eigenvalues of $L(t)$ correspond to the growth rate of $\delta x(t)$ in the corresponding directions. In the particular case represented by a constant matrix $\partial F/\partial x$, $L(t)$ can be diagonalized under the form

$$L(t) = \begin{vmatrix} e^{\lambda_1 t} & 0 & 0 \\ 0 & e^{\lambda_2 t} & 0 \\ 0 & 0 & e^{\lambda_3 t} \end{vmatrix}$$

The values λ_1, λ_2, and λ_3 are therefore the exponential growth rates along the eigen directions. When the matrix is not constant, the Lyapunov exponents can be found by looking for the equivalent λ_i when $t \to \infty$. These values are termed *characteristic* or *Lyapunov exponents*.

As we saw in Chapter 5, the dimension of an attractor can be fractal, which is important for chaotic motions as far as this dimension is linked to the number of the degrees of freedom of the system. A definition of this dimension rests on the definition of the Hausdorff–Besicovitch dimension, which we present first.

If we consider a Hilbert space H and a subensemble Y of H, for

$$d \in R^+ \quad \text{and} \quad \varepsilon > 0$$

we define

$$\mu(Y, d, \varepsilon) = \inf_I \sum_{i \in I} r_i^d$$

in which I is the minimum number of spheres (S) of diameter $r_i \le \varepsilon$ belonging to the family $(S_i)_{i \in I}$ necessary to cover Y. $\mu(Y, d, \varepsilon)$ is a decreasing function of ε, and μ is defined on $[0, \infty]$ by

$$\mu(Y, d) = \lim_{\varepsilon \to 0} \mu(Y, d, \varepsilon) = \sup_{\varepsilon > 0} \mu(Y, d, \varepsilon)$$

where μ is the measure of the Hausdorff dimension d of Y. It can be shown that there exists a number d_0 defined on $[0, \infty]$ such that

$$\mu(Y, d) = \infty \quad \text{when} \quad d < d_0$$

$$= 0 \quad \text{when} \quad d > d_0$$

where d_0 is the Hausdorff–Besicovitch dimension of the subspace Y.

$d_0 = d_H(Y)$, and we have

$$d_H(Y) = \inf\left\{\frac{d \in R^+}{\mu(Y, d)} = 0\right\}$$

since $d \rightarrow \mu(Y, d)$ is a positive decreasing function.

Note that Mandelbrot defined a Hausdorff–Besicovitch dimension in an analogous manner. Instead of choosing spheres of diameter d less than or equal to ε, he selected spheres of diameter equal to ε.

The fractal dimension of a system is defined by

$$d_f(Y) = \lim_{\varepsilon \to 0} - \frac{\ln n(\varepsilon)}{\ln \varepsilon}$$

in which $n(\varepsilon)$ is the minimum number of spheres of diameter less than ε necessary to cover Y.

Similarly, we can write

$$d_f(Y) = \inf\{d > 0, \mu_f(Y, d) = 0\}$$

$$\mu_f = \lim_{\varepsilon \to 0} \varepsilon^d n(\varepsilon)$$

since

$$\mu(Y, d) \leq \mu_f(Y, d)$$

we have $d_H \leq d_f$

For all ensembles found in R^n, the Hausdorff–Besicovitch, fractal, and topological dimensions are identical. For example, we saw in Chapter 5 that the snowflake (Koch figure) has a topological dimension of 1 but a fractal dimension $d_f = 1.26$; for the Lorenz attractor, the fractal dimension is 2.06. Note again that the fractal dimension is not necessarily a fraction. Indeed, the projection of the trajectories in Brownian motion has a topological dimension of 1, but because it is space filling, its fractal dimension is 2!

A finer measure of a strange attractor can be defined from a statistical point of view by using an information dimension that describes how the attractor is visited by the orbits. This information dimension is equal to the fractal dimension only for cases in which all the points of the attractor have the same probability of being traversed by the orbits. In general, these two dimensions are different.

If we consider an attractor "epsilon" and the corresponding number n of spheres of diameter ε necessary to cover it, and if P_i is the probability of the orbit passing through the sphere i, then we define the information I as

$$I(\varepsilon) = \sum_{i=1}^{n} -p_i \ln p_i$$

and the corresponding information dimension

$$d_I = \lim_{\varepsilon \to 0} -\frac{I(\varepsilon)}{\ln \varepsilon}$$

Because

$$\ln [n(\varepsilon)] > \ln I(\varepsilon)$$

we have $d_f > d_I$.

We can see that systems sensitive to initial conditions generate information because, by definition, two states initially infinitely close will be distinguishable after a finite time. The creation of information can be described by a new function called *entropy information*, which is null for predictable systems and positive for chaotic systems. This function has been shown by Ruelle to be always equal to or less than the sum of the Lyapunov exponents.

Finally, we mention the correlation dimension defined by Grassberger and Procaccia, which is presently enjoying much success because it enables easy adaptation for numerical studies. We define a correlation function as usual:

$$c(r) = \lim_{N \to \infty} \frac{1}{N^2} \sum_{i,j=1}^{N} \theta(r - \|\mathbf{x}_i - \mathbf{x}_j\|)$$

in which $\mathbf{x}_i = \mathbf{x}(t + i\tau)$, and θ is the heavyside function with τ an increment of time conveniently chosen. For a given r, $c(r)$ is the probability of finding two points of an orbit at a distance less than r. It can be shown that $c(r)$ varies as r^ν, where ν is the correlation dimension:

$$\nu = \lim_{r \to 0} \frac{\ln c(r)}{\ln \varepsilon}$$

Grassberger and Procaccia, with further work by Young, have established that $\nu = d_I \leq d_f$.

6.5 APPLICATION TO FLUID MECHANICS

Fluid mechanics rests on the Navier–Stokes equations, which are nonlinear partial differential equations, the types of equations that often describe chaotic systems. In classical fluid mechanics, we have always described turbulence as an evolution from a laminar state due to some instability of a perturbation. Only in the interpretation of a lower critical Reynolds number did we allude to an intrinsic flow property independent of such a disturbance. What is important in this chapter is that we now perceive that turbulence might not be due to an instability of a laminar state but to the very structure of the Navier–Stokes equations.

The formal study of systems of ordinary nonlinear differential (or difference) equations leads to the theory of chaotic systems. The Navier–Stokes equations, corresponding to an infinite number of ordinary equations because they are partial differential equations, are to be distinguished from usual systems of finite dimensions. The very existence of solutions of the Navier–Stokes equations has been addressed, but proved only for the two-dimensional case. For a viscous, incompressible steady flow in three dimensions, it has been shown that there are two critical Reynolds numbers, Re_1 and Re_2. When the flow is such that $Re < Re_1$, there is one solution, which is unique; when the Reynolds number lies between Re_1 and Re_2, there are several solutions; for Reynolds numbers greater than Re_2, no stationary solution exists. We should mention that the question as to whether two different initial states can lead to a unique solution has not yet been addressed.

For inviscid flow, since there is no dissipation, there is conservation of volumes in phase space; consequently, there is no attractor, and orbits are not confined to a limited domain. Such flows, described by the Euler equations, have all the characteristics of viscous flows except for the action of viscosity. Since inviscid flows are also nonlinear, we can expect chaos to be a feature of them. The role of viscosity, besides damping, is to limit the energy cascade to a certain scale, for which the velocity gradients are such that dissipation by the action of viscosity absorbs the energy that is passing through the spectrum. Notwithstanding the continuum assumption, the lack of viscosity corresponds to an infinite number of scales, so a global description of turbulence for inviscid flows is not possible.

As we have seen in Chapter 5, we elect to describe the flow in a lagrangian frame. In such a description, although Euler's equations give the evolution of the velocity field, the trajectories are obtained by integration of ordinary differential equations of the following form (see Chap. 5):

$$\dot{x} = u(x, y, z, t)$$
$$\dot{y} = v(x, y, z, t)$$
$$\dot{z} = w(x, y, z, t) \tag{6.6}$$

Moreover, if the flow is two dimensional, we have seen that the stream function Ψ plays the role of a hamiltonian. If the system is autonomous (independent of time), then we have an integrable system in which the trajectories correspond to the streamlines, and no chaos can appear. If the system is nonautonomous, then the particle trajectories move from streamlines to streamlines, and chaos can appear.

For viscous flows, we noted several experiments designed to elucidate possible routes to chaos; but on the theoretical side, much progress has recently been made, so theory is now ahead of experiment. Our hope here is to describe the transition to turbulence by means of a limited number of parameters. The observation that very simple nonlinear equations can lead to complex systems suggests that, conversely, problems as complex as turbulence could be described by very few parameters. Unfortunately, experiments suggest that the number of dimensions necessary to describe a turbulent flow might be high.

The Navier–Stokes equations have been the subject of several theoretical studies. It has been shown that for two-dimensional flows (for which we know that no turbulent regime can exist) a finite number of modes is sufficient to obtain solutions. More importantly, this result has now been extended to three-dimensional flows. If we consider the Kolmogorov length scale η to be representative of the scale below which viscous effects are predominant, then an order of magnitude of the number of degrees of freedom of the system contained in a cube of side L is N, given by $N = (L/\eta)^3$, where η is the Kolmogorov scale defined in terms of the kinematic viscosity ν and the rate of energy dissipation per unit mass ε:

$$\eta = \left(\frac{\nu^3}{\varepsilon}\right)^{1/4}$$

It then has been shown that the attractor on which the orbits evolve has a dimension less than or equal to D, given by $D = CN$. In the case of the two-dimensional mixing layer, we now consider the effects of viscosity, by writing the vorticity transport (Eq. 6.7) and the expression of the velocity in terms of the stream function (Eq. 6.8) and setting periodic boundary conditions. It is then natural to expand the stream function, as Lorentz did, in a double Fourier series, as shown by Eq. 6.9 where m and n are the spatial frequencies (modes) and the Ψ_{mn} are functions of time only:

$$(\Delta\Psi)_t + \Psi_x(\Delta\Psi)_y - \Psi_y(\Delta\Psi)_x = \nu\,\Delta^2\Psi \qquad (6.7)$$

$$u = -\frac{\partial\Psi}{\partial_y}(x,\,y,\,t) \qquad v = \frac{\partial\Psi}{\partial x}(x,\,y,\,t) \qquad (6.8)$$

After substitution in Eq. 6.7 and time normalization, we obtain a system of ordinary differential equations of the first order in which each Ψ_{mn} is governed by Eq. 6.10 and in which r is a normalized Reynolds number.

$$\Psi(x,\,y,\,t) = \sum_{m,n} \Psi_{mn}(t)e^{2\pi i(mx+ny)} \qquad (6.9)$$

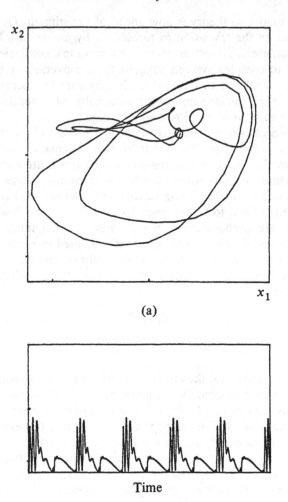

(a)

Time

(b)

Fig. 6.12   Periodic attractor. (a) Phase space, (b) time evolution. From same source as Fig. 5.1.

$$\dot{\Psi}_{mn}(t) + \frac{m^2 + n^2}{r}\Psi_{mn}(t)$$

$$= \sum_{p,q}(mq - np)\frac{p^2 + q^2}{m^2 + n^2}\Psi_{m-p,n-p}\Psi_{pq} \qquad (6.10)$$

If we select the eight following modes – (0, 1), (1, 1), (1, 0), (0, 2), (1, 2), (2, 2), (2, 1), and (2, 0) – we obtain a system of eight coupled ordinary differential equations that yields the following results:

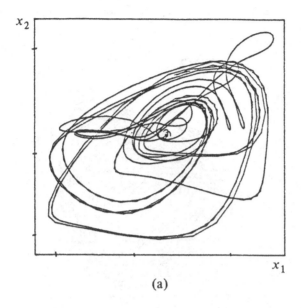

(a)

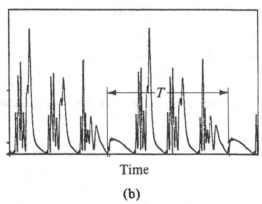

Time

(b)

Fig. 6.13   Period-doubling. (a) Phase space, (b) time evolution. From same source as Fig. 5.1.

$r < 0.45$	fixed point (steady case)
$r = 0.50$	periodic state, no decay (Fig. 6.12)
$r > 0.5311$	period doubling (Feingenbaum scenario)
	period attractor (Fig. 6.13)
$r > 0.5314$	strange attractor (Fig. 6.14)

The largest Lyapunov exponent corresponding to the strange attractor in Fig. 6.14 has in this case a value of 0.15 bits/s, as shown in Fig. 6.15. Which means that if we know the initial position on the attractor to

$x(t + \tau)$

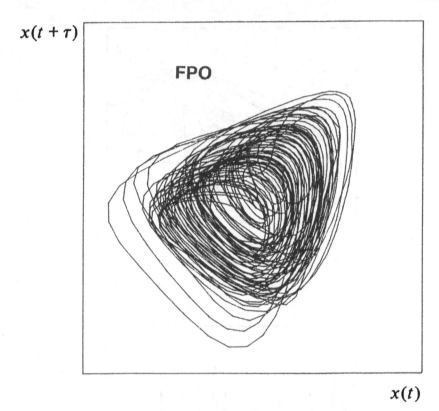

$x(t)$

Fig. 6.14   Computed strange attractor. From same source as Fig. 5.1.

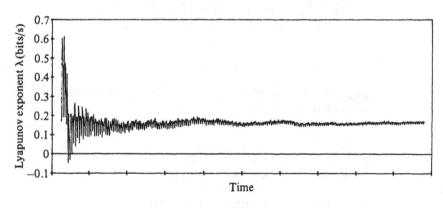

Fig. 6.15   Largest Lyapunov exponent in the shear layer. From same source as Fig. 5.1.

within one-millionth, prediction of its position after 153 is all but impossible because all the information is lost.

This last representation of the mixing layer, without simulating the fully turbulent flow, shows the clear advantage of a global description of the flow in which a streamline corresponds to a point. In this way, we can detect the presence of chaos within the Navier–Stokes equations describing the mixing layer.

Side view of a turbulent boundary layer. Streaklines from a smoke wire; Reynolds number 3500, based on momentum thickness. Photograph by T. Corke, Y. Guezennec, and H. Nagib. Reproduced by permission from M. van Dyke, *An Album of Fluid Motion*, the Parabolic Press, Stanford, California, 1982, p. 92.

7 Boundary Layer Theory

Physical problems become more and more complex as more effects interact, and the mathematical treatment becomes very difficult. Simplifications, however, can be expected when a single effect predominates. It is often possible to associate an operator with a typical physical role. By using classical vector operators, we can describe the part of a field that is locally associated with a rotating motion (curl operator) or the part of a field that is locally associated with a flow rate (divergence operator). Note that linear operators do not interact with one another.

If we consider a region of a space where both the curl and the divergence of the field are null, a scalar potential function Φ can be associated with this field, and we find that

$$\underbrace{\mathbf{V} = \text{grad } \Phi \qquad \text{div } \mathbf{V} = 0}_{\Delta\Phi = 0}$$

A diffusion phenomenon is likewise associated with a typical operator. At any point where the heat flux is conservative, the temperature function $T(x, t)$ satisfies the equation

$$\frac{\lambda}{\rho C} \frac{\partial^2 T}{\partial x_k \, \partial x_k} = \frac{\partial T}{\partial t} \qquad (7.1)$$

The physical properties of the conducting medium (λ, ρ, and C) are here assumed to be constant. The corresponding operator

$$\left\{ \frac{\partial}{\partial t} - \alpha \frac{\partial^2}{\partial x_k \, \partial x_k} \right\}$$

is therefore found in problems dealing with diffusion.

From this point of view, we can consider the Navier–Stokes equations

$$\underset{\text{II}}{\frac{\partial U_i}{\partial t}} + \underset{\text{I}}{U_j \frac{\partial U_i}{\partial x_j}} = \underset{\text{III}}{-\frac{1}{\rho} \frac{\partial P}{\partial x_i}} + \underset{\text{II}}{\frac{\mu}{\rho} \frac{\partial^2 U_i}{\partial x_k \, \partial x_k}} \qquad (7.2)$$

as describing complex mixtures of three effects: an advective effect (I) that is the result of the nonlinear operator $U_j \, \partial/\partial x_j$ when it is applied to the velocity; a diffusion process that is controlled by the kinematic viscosity (II); a pressure effect whose role will be described later (III); and an advective effect. All these effects are generally strongly inter-connected through the nonlinear character of the Navier–Stokes equations. Efforts have been made to separate these mechanisms by treating limiting cases. For instance, when the fluid velocity is very small at any point in a field, inertial terms can be neglected so that the initial problem turns into a new problem where diffusion effects predominate. If, on the other hand, the inertial forces dominate, the boundary layer theory as introduced by Prandtl can be used. The partial differential equation is still of the second order, but some of the terms of the highest degree drop out. The initial system of equations degenerates into another system. It is possible to postulate a theory for this case that is satisfactory insofar as attention is focused primarily on the physical aspects of this problem.

The concept of boundary layer was introduced by Prandtl at the beginning of this century in a somewhat heuristic manner. In this section, the initial concept of boundary layer is assumed to be known; we seek to extend this concept to other problems, which should be possible when some properties of a complex phenomenon dominate in one region of space and others dominate in another.

Briefly, the boundary layer hypothesis helps to bridge the gap between the intuitive expectation that the effects of viscosity on the flow are unimportant when the viscosity coefficient ν is small and when the no-slip condition must be satisfied at a solid boundary even though the viscosity coefficient is very small. This was Prandtl's main objective, a landmark in the development of fluid mechanics at the beginning of this century.

The role played in technical problems by the concept of boundary layer was most important, as has previously been mentioned, but the treatment of the Navier–Stokes equations has been developed in the last two decades in order to tackle the problems of complex flows that are made up of more or less extended regions of separate flows. The development of a flow with vorticity in the interior of the fluid can also be understood by considering the special case of a fluid that, being initially at rest, is set in motion by a solid body whose velocity rises suddenly at $t = 0$ from zero to some finite value and remains steady thereafter.

7.2 INTRODUCTION TO DEVELOPMENT OF FLOW WITH VORTICITY

A flow with vorticity develops when a solid body is given a finite velocity impulse and a fluid suddenly moves as a result. A brief study of this typical case supplies information about the evolution (at time $t > 0$) of the characteristic scales and the role of fundamental parameters at each stage of the development of the fluid motion.

First stage: The fluid motion at $t = 0$ is necessarily irrotational because the vorticity was zero at $t < 0$. The irrotational flow is fully specified by the known motion of the solid boundary. Thus, at $t = 0_+$ there is a sort of discontinuity in tangential velocity at the boundary, which is equivalent to a sheet vortex ω at the body surface. The line integral of ω along the normal to the boundary at any point is equal in magnitude to the local jump in tangential velocity, and is thus finite:

Second stage: The vorticity that was concentrated at the boundary at $t = 0_+$ diffuses into the fluid by the action of viscosity. If the changes in vorticity at a fixed point were due to viscous diffusion alone, the component of ω would satisfy the heat conduction equation, and the distance from the body to which vorticity would diffuse in time t would be of the order of magnitude $(\nu t)^{1/2}$. In fact, vorticity is also convected with a material element that produces a contribution to $\partial \omega / \partial t$ at a fixed point represented by the first term in the vortex equation. When complex flows are considered (three-dimensional flows), local distortions also come into play. In the case of a flat plate, these effects are eliminated. The velocity of the fluid relative to that of the body, at a position close to the body, has only a small normal component; so for small values of t, when the diffusion distance $(\nu t)^{1/2}$ is small, the main effect of advection is to convey vorticity parallel to the body surface rather than away from it. Thus, for small values of t, the vorticity of the fluid is not zero within a layer of thickness of order $(\nu t)^{1/2}$ surrounding the body. Within this layer the vorticity is finite because a finite velocity jump is now spread over a layer of nonzero thickness.

Third stage: During this third stage, $(\nu t)^{1/2}$ is no longer a small distance (compared with the characteristic length of the boundary), and advection is capable of conveying vorticity away from the body surface. As t goes to infinity, a steady motion of fluid relative to the body is established. The changes at a fixed point relative to the body that are due to advection by the fluid motion and to viscous diffusion then become null. Hence, the whole mechanism becomes a delicate balance between two mechanisms: an advective effect and a diffusive effect.

$$\underbrace{\frac{\partial \zeta}{\partial t} + \underbrace{U \frac{\partial \zeta}{\partial x} + V \frac{\partial \zeta}{\partial y}}_{\text{advective terms}} = \nu \left\{ \frac{\partial^2 \zeta}{\partial x^2} + \frac{\partial^2 \zeta}{\partial y^2} \right\}}_{\text{diffusive terms}} \tag{7.3}$$

In fact, the two operators that come into play are

$$\frac{\partial}{\partial t} - v\left\{\frac{\partial^2}{\partial x^2} + \frac{\partial^2}{\partial y^2}\right\} \quad \text{and} \quad U\frac{\partial}{\partial x} + V\frac{\partial}{\partial y}$$

which compete with each other; the flow field becomes steady when this competition turns into an equilibrium state.

First example: It is interesting to imagine some simple cases subjected to a single mechanism. In this first example, an infinite flat plate is given a finite velocity impulse, causing a fluid to move suddenly. The same velocity field is present at every point, so the unique component of the velocity is parallel to the plate; that is, parallel to ox. Consequently, we can write $V(y, t)$ with the single component $U(y, t)$. The inertial terms become null, and the nonlinear initial problem degenerates into a linear problem. The vortex field obeys a parabolic equation similar to that encountered in thermal conduction problems (see figure):

$$\frac{\partial \zeta}{\partial t} = \frac{\lambda}{\rho C}\frac{\partial^2 \zeta}{\partial y^2}$$

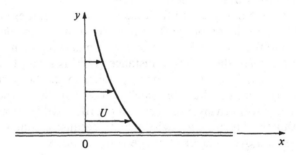

Second example: Now consider a wall made of porous material, such as sintered bronze or sheet metal perforated by many small holes, through which an adjustable amount of fluid can be sucked. An appropriate boundary condition is that the normal component of the relative velocity of fluid and solid at the boundary is equal to some assumed value that is determined by an expected property of the flow. The normal velocity at the wall is to be the same at all points of the boundary and equal to $-V$ (V being the suction velocity at the boundary). With regard to the condition to be satisfied by the tangential component of the relative velocity at the surface, experiment suggests that the no-slip condition is still appropriate; in any case, the values of the suction velocities ordinarily used in practice are much smaller than the main-flow velocities, so it should be reasonable to retain the no-slip condition at least as a fairly good approximation.

The solution given here first represents a steady two-dimensional flow over a plane solid boundary at which there is a uniform suction velocity V. The wall is assumed to be of such size (along ox) that we can postulate that the flow variables are independent of position in any plane parallel to the boundary. It is possible thereby to eliminate the inertial terms, as was done in the first example. By using the continuity equation we can verify that the following data are self-consistent: x, y, and z are rectilinear coordinates with y normal to the boundary, the fluid velocity is $(U, -V, 0)$ and the vorticity is $(0, 0, \zeta)$ where $\zeta = -dU/dy$. The vorticity equation becomes

$$-V\frac{d\zeta}{dy} = \nu \frac{d^2\zeta}{\partial y^2}$$

or upon integration,

$$V(\zeta_0 - \zeta) = \nu \frac{d\zeta}{dy} \tag{7.4}$$

which simply says that the rate at which (the excess of) vorticity is conveyed across the unit area of the (x, z) plane by convection with velocity $-V$ (see figure), exactly cancels the rate of transport of viscous diffusion. One more integration yields

$$\zeta - \zeta_0 = -\frac{dU}{dy} - \zeta_0 = Ae^{-Vy/\nu}$$

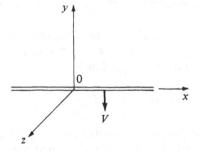

The region of nonuniform vorticity extends to a distance of order ν/V from the boundary, which is to be expected from the general arguments about the balance between convection toward the boundary and viscous diffusion away from it.

7.3   BOUNDARY LAYER APPROXIMATION

For the basic features of this section we refer to the previous paragraphs. However, we recall here the order-of-magnitude treatment of the terms in the Navier–Stokes equations when the Reynolds number goes to infinity and the region of flow disturbed by viscous effects becomes smaller and smaller.

Steady two-dimensional flows of viscous, incompressible fluids are governed by the following equations:

$$U \frac{\partial U}{\partial x} + V \frac{\partial U}{\partial y} = -\frac{1}{\rho} \frac{\partial P}{\partial x} + \nu \frac{\partial^2 U}{\partial x^2} + \nu \frac{\partial^2 U}{\partial y^2} \qquad (7.5)$$

$$U \frac{\partial V}{\partial x} + V \frac{\partial V}{\partial y} = -\frac{1}{\rho} \frac{\partial P}{\partial y} + \nu \frac{\partial^2 V}{\partial x^2} + \nu \frac{\partial^2 V}{\partial y^2} \qquad (7.6)$$

$$\frac{\partial U}{\partial x} + \frac{\partial V}{\partial y} = 0 \qquad (7.7)$$

The flow is assumed to develop with a preferential direction (along *ox*). For the purpose of explaining these approximations, we take the boundary to be a plane wall; however, all subsequent approximations can be made if the flow under consideration exhibits such a preferential direction. Hence, this theory is applicable to jets, wakes, mixing zones, and so on. The boundary layer can be considered as the zone where the diffusion of vorticity by viscous effects is significant compared with advective effects. The boundary layer thickness (defined in a suitable way) is assumed to be small everywhere compared with the characteristic length in the flow direction over which the flow velocity changes appreciably. L denotes a characteristic dimension in the direction of the flow; l is the corresponding dimension normal to the wall. As was previously said, $L \gg l$.

The flow velocity outside the boundary layer is denoted by U_0, and the order of magnitude of $\partial U/\partial x$ is $\partial U/\partial x \sim U_0/L$. We infer the order of magnitude of V from the continuity equation

$$V_\delta = -\int_0^\delta \frac{\partial U}{\partial x} \, dy \Rightarrow V \sim U_0 \frac{l}{L}$$

In so doing, we overestimate the order of magnitude of V. An assessment of the order of magnitude of all the terms in the momentum equation becomes possible except for the pressure term, which will be evaluated later. The x-component of the Navier–Stokes equations is

$$\underbrace{U \frac{\partial U}{\partial x}}_{U_0^2 L^{-1}} + \underbrace{V \frac{\partial U}{\partial y}}_{U_0^2 L^{-1}} = -\frac{1}{\rho} \frac{\partial P}{\partial x} + \underbrace{\nu \frac{\partial^2 U}{\partial x^2}}_{\frac{\nu}{U_0 L} U_0^2 L^{-1}} + \underbrace{\nu \frac{\partial^2 U}{\partial y^2}}_{\frac{\nu}{U_0 L} U_0^2 L^{-1} \frac{L^2}{l^2}}$$

The two inertial terms have the same order of magnitude whereas the two viscous terms are very different from each other. Inasmuch as we must account for viscous effects in all cases, it is reasonable to assume that the largest viscous term is of the same order of magnitude as the inertial terms. This leads to

$$\frac{\nu}{U_0 L} \frac{L^2}{l^2} \sim 1$$

that is, $Re_L \sim \dfrac{L^2}{l^2}$

Accordingly, this theory is applicable to a certain range of Reynolds numbers; in fact, it is for high Reynolds numbers only. Note that for consistency, $\partial P/\partial y$ should be of the same order of magnitude as the inertial term.

The second component of the momentum equation can be written and an order of magnitude analysis made:

$$U\frac{\partial V}{\partial x} + V\frac{\partial V}{\partial y} = -\frac{1}{\rho}\frac{\partial p}{\partial y} + \nu\frac{\partial^2 V}{\partial x^2} + \nu\frac{\partial^2 V}{\partial y^2}$$

$$\underbrace{U_0^2 L^{-1}\frac{l}{L}} \quad \underbrace{U_0^2 L^{-1}\frac{l}{L}} \qquad \qquad \underbrace{\frac{\nu}{U_0 L}U_0^2 L^{-1}\frac{l}{L}} \quad \underbrace{\frac{\nu}{U_0 L}U_0^2 L^{-1}\frac{L}{l}}$$

We use here the order of magnitude $U_0^2 L^{-1}$ previously encountered for making comparisons. The order of magnitude of all the terms in this equation is equal to or inferior to ε. Inertial terms are of an order of magnitude $l/L \sim$ ε smaller than those encountered in the previous equation. The second viscous term is also of the same order of magnitude if set as before:

$$\frac{\nu}{U_0 L} \cdot \frac{L^2}{l^2} \sim 1$$

The first viscous term is obviously negligible. The only term whose order of magnitude is still unknown, $\partial P/\partial y$, should also be of an order of magnitude equal to ε. Accordingly, $\partial P/\partial y \sim$ ε and P depends only on x, and at any point x in the boundary layer the pressure is constant and equal to the pressure in the inviscid fluid, just outside the boundary layer:

$$P(x) = P_\delta(x)$$

Therefore, the pressure is not an unknown function: The pressure field can be extracted from a previous iteration (computation available in the case of an inviscid fluid).

To a large extent, the new role of the pressure is determinant in the boundary layer theory; this evolution is far more important than the dominant role of the second viscous term with respect to the first one.

That will become clear when the evolution of the mathematical properties of the equations of motion is discussed.

7.4 THE MATHEMATICAL STRUCTURE OF THE EQUATIONS

The mathematical nature of the problem is strongly modified according to whether the Navier–Stokes equations or the boundary layer equations are considered. A basic view of the problems as they relate to systems of partial differential equations is briefly introduced in Appendix C. In the first stage we prefer to deal with the Euler equation in order not to obliterate the mathematical character of the equations through the second derivatives, whose role is determinant as far as a mathematical classification is concerned. In other words, the two components F_x and F_y of the stress forces are assumed to be known quantities. Accordingly, the set of equations under consideration can be written

$$U \frac{\partial U}{\partial x} + V \frac{\partial U}{\partial y} = -\frac{1}{\rho} \frac{\partial P}{\partial x} + Fx \qquad \text{(known)}$$

$$U \frac{\partial V}{\partial x} + V \frac{\partial V}{\partial y} = -\frac{1}{\rho} \frac{\partial P}{\partial y} + Fy \qquad \text{(known)}$$

$$\frac{\partial U}{\partial x} + \frac{\partial V}{\partial y} = 0$$

These equations, in matrix form, are

$$\begin{bmatrix} U & 0 & 1/\rho \\ 0 & U & 0 \\ 1 & 0 & 0 \end{bmatrix} \begin{bmatrix} \partial U/\partial x \\ \partial V/\partial x \\ \partial P/\partial x \end{bmatrix} + \begin{bmatrix} V & 0 & 0 \\ 0 & V & 1/\rho \\ 0 & 1 & 0 \end{bmatrix} \begin{bmatrix} \partial U/\partial y \\ \partial V/\partial y \\ \partial P/\partial y \end{bmatrix} = \begin{bmatrix} F_x \\ F_y \\ 0 \end{bmatrix}$$

We assume that the usual theories of partial differential equations are known by the reader, so we can now determine the characteristics of this differential system.

The associated determinant can be written as

$$\begin{vmatrix} U - \lambda V & 0 & 1/\rho \\ 0 & U - \lambda V & -\lambda/\rho \\ 1 & -\lambda & 0 \end{vmatrix} = -\frac{1}{\rho}[1 + \lambda^2][U - \lambda V] \qquad (7.8)$$

so the characteristic directions at a given point are given by the equation

$$(1 + \lambda^2)(U - \lambda V) = 0 \qquad (7.9)$$

Two characteristics associated with the system are imaginary, the third is real and corresponds to a streamline. This complex system exhibits an elliptic character through the two imaginary roots of the previous λ equation and simultaneously a hyperbolic character through the real root of the λ equation. This characteristic is a streamline, which means that any disturbance can be conveyed by the velocity field along a streamline. This finding is consistent with physical concepts.

The elliptic character of the differential system reinforces the role of the pressure terms, which are assumed to be unknown quantities. Due to the pressure terms, all parts of the flow field are strongly interdependent, so the flow must be considered as a whole. It is this that renders the problem so complex. Previously, this interplay among all the parts of the field was displayed by a Poisson equation. We recall that the momentum equation, supplemented by the continuity equation, leads to

$$\frac{\partial U_i}{\partial t} + U_j \frac{\partial U_i}{\partial x_j} = -\frac{1}{\rho}\frac{\partial P}{\partial x_i} + \nu \frac{\partial^2 U_i}{\partial x_j\,\partial x_j} \qquad (7.10)$$

which for an incompressible fluid can be written as

$$\frac{\partial P}{\partial x_i \partial x_i} = -\rho \frac{\partial^2 U_i U_j}{\partial x_i \partial x_j}$$

This Poisson equation has the following particular solution,

$$P(\mathbf{x}) = -\frac{1}{4\pi} \iiint \rho \frac{\partial^2 U_i U_j(\mathbf{x}')}{\partial x_i \partial x_j}\,d\mathbf{x}' \qquad (7.11)$$

which, if the simplified equations are considered, becomes

$$U \frac{\partial U}{\partial x} + V \frac{\partial U}{\partial y} = -\frac{1}{\rho}\frac{\partial P}{\partial x} + F_x$$

$$\frac{\partial U}{\partial x} + \frac{\partial V}{\partial y} = 0$$

The x component of the momentum equation is supplemented by the continuity equation; the characteristics of this system are given by the determinant Δ that is associated with the matrix form of the equations:

$$\Delta = \begin{vmatrix} U - \lambda V & 0 \\ 1 & -\lambda \end{vmatrix} = -\lambda[U - \lambda V] = 0$$

This gives two real characteristics:

$$\lambda = 0 \Rightarrow \frac{dy}{dx} = \infty$$

$$\lambda = \frac{U}{V} \Rightarrow \frac{dy}{dx} = \frac{V}{U}$$

In the boundary layer approximation, the pressure becomes a known function and the mathematical role of this function vanishes. Streamlines are still characteristic lines of this system, the second characteristics being parallel to Oy, which is normal to the wall (Fig. 7.1). It is a direct consequence of the equation $\partial p/\partial y = 0$. Information linked to the pressure is immediately transmitted along a line parallel to Oy.

Characteristic lines are preferential pathways along which information is conveyed. The initial system having a unique character, the number of real characteristics becomes equal to the number of unknown equations considered.

In as much as information goes along characteristic lines,* their role is essential when wave problems are considered.

This problem has to be reexamined if viscous effects are taken into account. We now consider a two-dimensional flow described by

$$U \frac{\partial U}{\partial x} + V \frac{\partial U}{\partial y} = -\frac{1}{\rho} \frac{\partial P}{\partial x} + \nu \left[\frac{\partial^2 U}{\partial x^2} + \frac{\partial^2 U}{\partial y^2} \right]$$

$$\frac{\partial U}{\partial x} + \frac{\partial V}{\partial y} = 0$$

in which the terms of highest order are the viscous terms; accordingly, the first equation has an elliptic character. If the Prandtl approximation is made, we have

$$\frac{\partial^2 U}{\partial y^2} \gg \frac{\partial^2 U}{\partial x^2}$$

and thus,

$$U \frac{\partial U}{\partial x} + V \frac{\partial U}{\partial y} = -\frac{1}{\rho} \frac{\partial P}{\partial x} + \nu \frac{\partial^2 U}{\partial y^2} \tag{7.12}$$

This is a partial differential equation of second order with respect to the independent variable y and of first order with respect to the independent variable x. It can be compared directly with the thermal equation

$$\frac{\partial T}{\partial t} = \frac{\lambda}{\rho C} \frac{\partial^2 T}{\partial x^2} \tag{7.13}$$

which is second order in x and first order in t.

---

* In the boundary layer approximation, the fluid is assumed to be incompressible, and accordingly, the traveling time for any disturbance is null (the "velocity wave" being equal to infinity). With a limited time of propagation, a time delay has to be introduced; information collected in x at time t started from x' at time $t - \theta$. The speed of propagation is termed C.

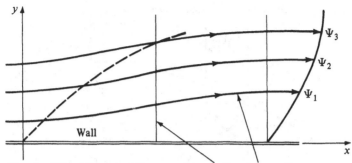

Two families of characteristics: normal to wall and streamline

Fig. 7.1   Characteristics in a boundary layer

With a Prandtl approximation, the role of pressure acting as an unknown function has been eliminated, and one of the viscous terms is seen to dominate the other one. The final equation now becomes a parabolic equation.

The nonlinear character is held by the advective terms. This equation must be associated with the following boundary conditions:

$$U(x, 0) = V(x, 0) = 0 \qquad \text{no-slip condition at the wall}$$
$$U(x, \infty) = U_0(x) \qquad \text{matching condition}$$
$$U(0, y) = U(y) \qquad \text{initial condition}$$

The parabolic character of the equation is emphasized if x and Ψ (stream function) are taken as independent variables (in place of x and y). This change of variables is possible if a one-to-one relation exists between the two variables y and Ψ, which means that no separation point exists in the region under consideration (Fig. 7.2).

$$\left.\frac{\partial}{\partial x}\right|_y = \left.\frac{\partial}{\partial x}\right|_\Psi - \left.\frac{\partial}{\partial \Psi}\right|_x V$$

$$\left.\frac{\partial}{\partial y}\right|_x = U \left.\frac{\partial}{\partial \Psi}\right|_x$$

The momentum equation can be written as

$$U \frac{\partial U}{\partial x} = -\frac{1}{\rho} \frac{\partial P}{\partial x} + \nu U \frac{\partial}{\partial \Psi}\left[U \frac{\partial U}{\partial \Psi}\right]$$

and if we let

$$f = \frac{U^2 - U_0^2}{2} = \frac{U^2}{2} + \frac{P}{\rho} + C \Rightarrow U^2 = 2f + U_0^2$$

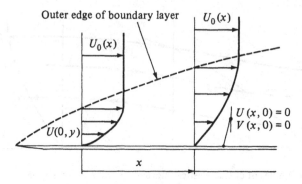

Fig. 7.2   Boundary layer development

it finally becomes

$$\frac{\partial f}{\partial x} = \nu\sqrt{2f + U_0^2}\,\frac{\partial^2 f}{\partial \Psi^2} \qquad (7.14)$$

with the following new boundary conditions:

$$f(x, 0) = \frac{P}{\rho}$$

$$f(x, \infty) = \frac{P}{\rho} + \frac{U_0^2}{2} = \text{constant}$$

$$f(0, \Psi) = \frac{P(0)}{\rho} + \frac{U^2(\Psi)}{2}$$

We recall that the stream function is linked to the velocity field by the two relations

$$U = \frac{\partial \Psi}{\partial y} \qquad \text{and} \qquad V = -\frac{\partial \Psi}{\partial x}$$

so by using a stream function, we verify the continuity equation automatically. Reexamining the validity of the previous transformation, its jacobian must be evaluated

$$J = \frac{D(x, \Psi)}{D(x, y)} = U$$

Its sign obviously depends on the sign of U. A change of sign of this jacobian is linked to the existence of a separation point with a reverse motion (Fig. 7.3). Through these reverse motions, information is conveyed in all directions (downstream to upstream); in the wall region the flow is influenced everywhere by the shape of the whole boundary of

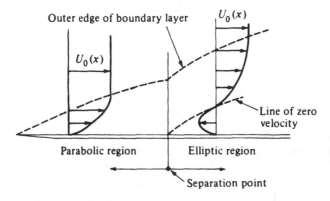

Fig. 7.3 Separation

this region of flow, including that part of the boundary within the separation region. In fact, in the region of flow governed by a parabolic equation, the flow develops everywhere in the same direction (x), whereas in the region of reverse flow, the flow is governed by an elliptic equation and a preferential direction of flow does not exist. In this case, the change of variable $(x, y$ to $x, \Psi)$ fails.

Disregarding the nonlinearity of the equation with x and Ψ as independent variables, we see that the main difficulty is the existence of a singularity located at the wall (separation point). The physical nature of the problem depends on the region under consideration as well as the mathematical nature of the associated equations.

7.5 AN INTRODUCTION TO SINGULAR PERTURBATION

The mathematical nature of a set of partial differential equations such as the Navier–Stokes equations depends on the terms of highest degree. By neglecting some of these terms we can strongly alter the nature of the problem. Passing from the Navier–Stokes equations to the boundary layer equation gives rise to fundamental changes; for instance, not all flows exhibiting recirculation zones can be considered. It is still possible to satisfy the no-slip condition when we deal with problems having a preferential direction. Even though the viscosity of the fluid becomes very small, this no-slip condition can be verified. If the second-order terms are dropped, the Navier–Stokes equations become the Euler equations, and the problem is drastically simplified.

We wonder what happens when the terms of highest degree in an equation become smaller and smaller without becoming null. This

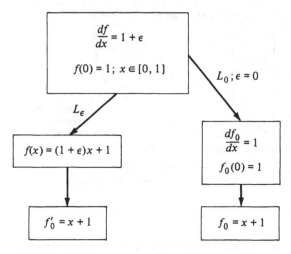

Fig. 7.4   Regular perturbation scheme

situation can be duplicated by controlling these terms with a parameter that tends to zero without ever becoming null. The approach to zero of this small parameter should generate strong modifications, including a fundamental revision of the boundary conditions. In fact this difficulty gives rise to fundamental considerations concerning the handling of physical problems. When treating a physical problem, we emphasize the role of some features and disregard others. Even though certain secondary phenomena intervene very weakly in the equations, they can dominate when they are represented by terms of highest degree in the equation. If we neglect secondary phenomena, we first assume that they appear in the equations through small terms; we also admit that they do not interfere with the highest-degree term in the equation. In other words, this cancelation is assumed not to introduce a *singular perturbation*.

7.5.a   *Two simple cases*

Before examining boundary layers in particular, let us look at two simple cases.

Regular perturbations, the case of ordinary differential equations: We consider two differential operators: L_ε, the complete operator including the small terms (of order of magnitude ε) and L_0, a degenerate operator that is obtained by setting $\varepsilon = 0$ in the equation itself. The solution of the complete equation includes terms of the order of magnitude ε, and we ask ourselves what happens when ε goes to zero in

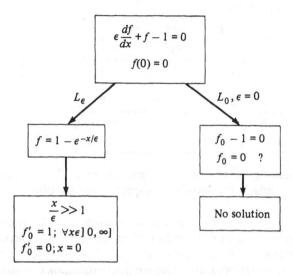

Fig. 7.5   Singular perturbation. A solution for f_0 does not exist.

this solution? The solution of the truncated equation has no term of the order of magnitude ε. What is this solution in comparison with the one obtained with L_0 (Fig. 7.4)?

> The solution of the complete equation f is called f_0' when $\varepsilon \to 0$ whereas the solution of the truncated equation is termed f_0. It can be seen that f_0' is identical to $f_0 : f_0' \equiv f_0$.

Singualr perturbation: We treat an ordinary differential equation whose differential term depends on ε as in Fig. 7.5.

7.5.b   *Partial differential equations*

Regular perturbation: Now we examine a Poisson equation whose second term is of an order of magnitude ε.

$$\frac{\partial^2 \Phi}{\partial x^2} + \frac{\partial^2 \Phi}{\partial y^2} = \varepsilon(x, y)$$

solution $L_\varepsilon(\varepsilon = 0)$ $\longrightarrow$ solution L_0

$$L_\varepsilon(\varepsilon = 0) \to L_0$$

Singular perturbation:

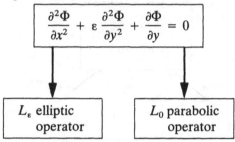

The initial operator degenerates as ε goes to zero. With a parabolic operator, the problem develops with respect to the variable y.

A matching process: Always, when working with an ordinary differential equation, we can introduce a matching method through an expanded operator $\hat{L}$. This operator is also called an inner operator, whereas L_0 is an outer operator. The complete operator consists of an outer and an inner operator. This method can be interpreted as the introduction of an adequate scaling in direct connection with the order of magnitude of the small terms encountered in the equation. Dealing with a space variable means that we will reexamine the problem through new variables. We will use η in place of x and set $x = \varepsilon\eta$ in which η is the inner variable.

Such a method can be applied to temporal phenomena, in which case an expanded time $t = \varepsilon\tau$ is introduced instead of an extended length. The expanded time is self-adjustable when ε is introduced in the change of t. The general scheme is presented in Fig. 7.6.

The next example starts from an ordinary differential equation. The complete solution is presented and the fundamental approximation given in Fig. 7.7. The degenerate operator is also considered, and the expanded (or inner) operator will be introduced later. The operator can be expanded in accordance with the small parameter:

$$x = \varepsilon\eta \Rightarrow \hat{L}(\eta)$$

This change of variable leads to

$$\frac{d^2f}{d\eta^2} + \frac{df}{d\eta} = 0 \tag{7.15}$$

and the terms become of the same order of magnitude.

The solution of this equation, which verifies the boundary condition α, is termed $\hat{f}_\alpha$ and can be written as

$$\hat{f}_\alpha = (1 - a)[1 - e^{-\eta}]$$

For $\eta \Rightarrow \infty$, $\hat{f}_\alpha = 1 - a$, the value found for f_0 with $x = 0$.

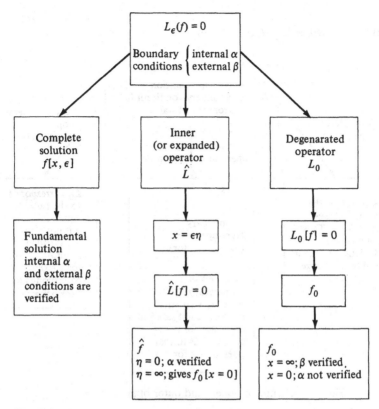

Fig. 7.6   General scheme of the matching process

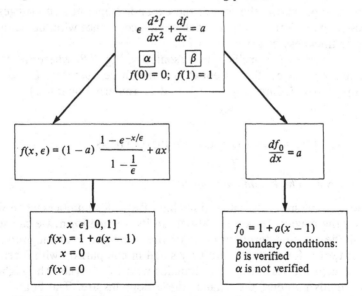

Fig. 7.7   Complete solution and approximation of ordinary differential equation

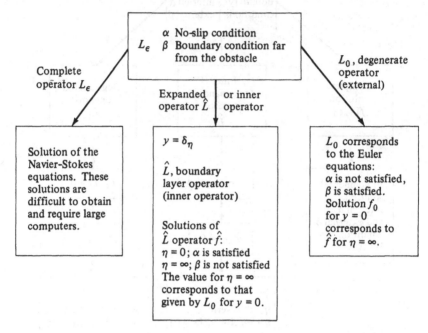

Fig. 7.8   Complete, inner, and outer operators

This example shows the role of the expanded operator and suggests the possibility of introducing a variable η in accordance with the boundary layer thickness by setting $y = \delta\eta$.

An overview of the problem is presented in Fig. 7.8, where we consider, for the Navier–Stokes equations, the three operators L_ε (complete operator), $\hat{L}$ (inner operator), and L_0 (external operator).

7.6   THE BOUNDARY LAYER EQUATIONS

7.6.a   *The kinematic problem*

A similar problem is encountered for high Reynolds numbers when we consider the Navier–Stokes equations. In its original form, we can see that all the existing terms are not of the same order of magnitude. Viscous terms, for example, are very small in comparison with inertial terms, except for very restricted domains next to the wall. The higher the Reynolds number, the smaller these domains are. The boundary layer can be considered to be a matching zone in which the behavior of an inviscid fluid is readapted to that of a viscid fluid with no-slip conditions at the walls. In this zone, the reduction of layer thickness steepens the concentration gradients.

We shall limit our approach to two-dimensional flows, so the set of equations that we have to consider are

$$U\frac{\partial U}{\partial x} + V\frac{\partial U}{\partial y} = -\frac{1}{\rho}\frac{\partial P}{\partial x} + \nu\left[\frac{\partial^2 U}{\partial x^2} + \frac{\partial^2 U}{\partial y^2}\right]$$

$$U\frac{\partial V}{\partial x} + V\frac{\partial V}{\partial y} = -\frac{1}{\rho}\frac{\partial P}{\partial y} + \nu\left[\frac{\partial^2 V}{\partial x^2} + \frac{\partial^2 V}{\partial y^2}\right]$$

$$\frac{\partial U}{\partial x} + \frac{\partial V}{\partial y} = 0$$

The second equations can be rewritten as

$$\frac{1}{\rho}\frac{\partial P}{\partial y} = -U^2\frac{\partial}{\partial x}\left[\frac{V}{U}\right] - \nu\left[\frac{\partial^2 V}{\partial x^2} + \frac{\partial^2 V}{\partial y^2}\right]$$

and the boundary layer approximation yields

$$V \sim \varepsilon$$

and therefore

$$\frac{1}{\rho}\frac{\partial P}{\partial y} \sim \varepsilon \Rightarrow P = P(x)$$

The two following equations are then sufficient to treat this problem:

$$U\frac{\partial U}{\partial x} + V\frac{\partial U}{\partial y} = -\frac{1}{\rho}\frac{\partial P}{\partial x} + \nu\frac{\partial^2 U}{\partial y^2}$$

$$\frac{\partial U}{\partial x} + \frac{\partial V}{\partial y} = 0$$

To set the problem in an expanded form, we must introduce a characteristic scale $\beta(x)$. This scale must be closely related to the viscous term included in the momentum equations. Doing so, we can give the accurate form taken by the inner operator $\hat{L}$. We set

$$y = \eta\beta(x) \qquad \frac{U(x, y)}{U_0(x)} = f(\eta, x) \tag{7.16}$$

In this form, the role of two scales is emphasized: a *length scale* $\beta(x)$, which permits an expansion in the y direction, and a *velocity scale* $U_0(x)$, which is (for example) the velocity just outside the boundary layer (it is also the velocity at the wall given by the treatment of the operator L_0 corresponding to the Euler equations). If the inviscid flow is assumed to be irrotational, the velocity at the wall is given by the resolution of the potential equation).

It is easy to hold the dependence of $U_0(x, y)/U_0(x)$ on x. To simplify the algebraic treatment, we assume an *equilibrium state*, which means that the problem is assumed to be frozen with respect to the x variable through a convenient choice of the two scales $\beta(x)$ and $U_0(x)$. In this way, we extract a restricted set of solutions of the boundary layer equations, but the initial problem of a matching process is not essentially modified.

If $U_0(x)$ and $\beta(x)$ are the two characteristics scales, the most obvious choice would be to take $U_0(x)$ equal to the velocity just outside the boundary layer in the region where viscous effects do not intervene, and $\beta(x)$ equal to the boundary layer thickness. At this stage, it is better to maintain flexibility, as will be shown later.

If an equilibrium state exists, then we can write

$$\frac{U(x, y)}{U_0(x)} = f(\eta)$$

which, after the change of variables, becomes

$$\frac{\partial U}{\partial x} = U_0'f - U_0 \frac{\beta'}{\beta} \eta f' \tag{7.17}$$

$$U \frac{\partial U}{\partial x} = U_0 U_0' f^2 - U_0^2 \frac{\beta'}{\beta} \eta f f' \tag{7.18}$$

$$U = U_0(x) \quad\text{and}\quad f = f(\eta)$$

where the primes for U_0 denote derivatives with respect to x and for f with respect to η.

The velocity component V can be eliminated through the continuity equation. And because at the wall, $V = 0$ for $y = 0$ (no suction, no blowing), we can write

$$V(y) = -\int_0^y \frac{\partial U}{\partial x} \, dy$$

$$V(\eta) = -\beta U_0' \int_0^\eta f(\eta) \, d\eta + \beta' U_0 \int_0^\eta \eta f' \, d\eta$$

We now introduce a function $F(\eta)$, related to $f(\eta)$, that plays a role similar to that of a dimensionless stream function.

$$\int_0^\eta f(\eta) \, d\eta = F(\eta) \tag{7.19}$$

$$f(\eta) = \frac{U(\eta)}{U_0} = \frac{dF}{d\eta} = F' \tag{7.20}$$

so

$$V(\eta) = -\beta U_0' F + U_0 \beta' \int_0^\eta \eta f' \, d\eta$$

and

$$V \frac{\partial U}{\partial y} = -U_0 U_0' f' F + U_0^2 \frac{\beta'}{\beta} f' \int_0^\eta \eta f' \, d\eta$$

The inertial terms can be rearranged to obtain

$$U \frac{\partial U}{\partial x} + V \frac{\partial U}{\partial y} = U_0 U_0' f^2 - U_0 U_0' f' F$$

$$+ \frac{U_0^2 \beta'}{\beta} \left[-ff'\eta + f' \int_0^\eta \eta f' \, d\eta \right] \quad (7.21)$$

in which the last integral in square brackets can be expressed as

$$f' \int_0^\eta \eta f' \, d\eta = f'[\eta f - F]$$

and the inertial terms as

$$U \frac{\partial U}{\partial x} + V \frac{\partial U}{\partial y} = U_0 U_0' f^2 - U_0 U_0' f' F + U_0^2 \frac{\beta'}{\beta} f' F$$

The motion outside the boundary layer being assumed irrotational, the constant of the motion yields

First choice velocity scale $\begin{cases} P + \dfrac{1}{2} \rho U_0^2 = \text{constant} \\ -\dfrac{1}{\rho} \dfrac{\partial P}{\partial x} = U_0 U_0' \end{cases}$

In this form of the constant of the motion, we make a choice regarding the velocity scale, which is taken equal to U_0, the velocity just outside the boundary layer. This boundary layer is again assumed to be very thin. The viscous terms can be written

$$\frac{\partial U}{\partial y} = \frac{U_0 f'}{\beta} \qquad \tau = \mu \frac{\partial U}{\partial y} = \mu \frac{U_0}{\beta} f'$$

$$\nu \frac{\partial^2 U}{\partial y^2} = \frac{\nu U_0}{\beta^2} f'' \qquad \frac{\tau}{\frac{1}{2}\rho U_0^2} = \frac{\nu}{U_0 \beta} f'$$

Finally, the boundary layer equation becomes

$$U_0 U_0' f^2 - U_0 U_0' f' F + \frac{U_0^2 \beta'}{\beta} f' F = U_0 U_0' + \nu \frac{U_0}{\beta^2} f'' \quad (7.22)$$

and by multiplying by $\beta/U_0^2\beta'$, we obtain an equation relative to the inner operator $\hat{L}$:

$$\frac{\beta U_0'}{\beta' U_0}[(F')^2 - F''F] - F''F = \frac{\beta U_0'}{\beta' U_0} + \frac{\nu}{U_0\beta\beta'} F''' \qquad (7.23)$$

Starting with the Navier–Stokes equations and considering only two-dimensional flows, we introduce a streamline function: the final equation for Ψ, which is of the fourth degree. In the case under consideration, the equation for F (which is equivalent to Ψ) is only third order because of the pressure terms, which are unknown in the Navier–Stokes equations but are known functions in the boundary layer approximation.

The previous equations depend on x through the two terms

$$\frac{\beta U_0'}{\beta' U_0} \quad \text{and} \quad \frac{\nu}{U_0\beta\beta'}$$

If an equilibrium state exists, this equation becomes independent of x, and the following two terms must be constants:

$$\frac{\beta U_0'}{\beta' U_0} = b \qquad \frac{\nu}{U_0\beta\beta'} = a$$

The previous equation can now be written as

$$b[(F')^2 - FF''] - F''F = b + aF'''$$

Therefore, the two characteristic scales are closely related:

$$\frac{U_0'}{U_0} = b\frac{\beta'}{\beta} \Rightarrow U_0 = C\beta^b$$

U_0 can be eliminated by using the second relation:

$$\frac{\nu}{C\beta^{b+1}\beta'} = a \Rightarrow \beta = \left[\frac{b+2}{Ca}\nu x\right]^{1/(b+2)} \sim xp$$

No constant of integration has been introduced, which is tantamount to assuming that the origin of the axes coincides with that of the boundary layer.

U_0 can be presented as a function of x:

$$U_0 = C\beta^b = C\left[\frac{b+2}{Ca}\nu x\right]^{b/(b+2)}$$

$$U_0 = \left[C\frac{b+2}{Ca}\nu\right]^{b/(b+2)} x^{b/(b+2)}$$

Setting

$$m = \frac{b}{b + 2} \Rightarrow b = \frac{2m}{1 - m}$$

we write

$$U_0 \sim x^m$$

The external distribution of the velocity corresponds to a power law.
 The choice of the length scale can now be made by taking into account:

$$\beta^{b+2} = \frac{b + 2}{a} \frac{vx}{C}$$

$$\beta^2 = \frac{b + 2}{a} \frac{vx}{C\beta^b} = \frac{b + 2}{a} \frac{vx}{U_0}$$

The writing becomes simpler by setting $(b + 2)/a = 1$. Accordingly,
the scaling is completed by choosing the length scale, to be given by

$$\beta = \sqrt{\frac{vx}{U_0}}$$

Thus, the fundamental equation, simplified by $b = 2m/(1 - m)$ and
$a = 2/(1 - m)$, finally becomes

$$F''' + \frac{1 + m}{2} FF'' - m(F')^2 = 0$$

In the equation

$$\frac{\beta U_0'}{\beta' U_0} [(F')^2 - F''F] - F''F = \frac{\beta U_0'}{\beta' U_0} + \frac{v}{U_0 \beta \beta'} F''' \qquad (7.24)$$

which corresponds to the inner operator, all the terms are of the same
order of magnitude. This property characterizes the operator $\hat{L}$. The
previous equation contains a set of equilibrium profiles that is important
to initiate a boundary layer computation.

7.6.b *Boundary layer developing along a flat plate*
     (Blasius's solution)

The characteristic velocity has been chosen just outside the boundary
layer, so it becomes

$$U_0 \sim x^m \Rightarrow m = 0$$

and the equation takes on the simple form

$$2F''' + FF'' = 0 \qquad (7.25)$$

The length scale η is equal to

$$\eta = \frac{y}{\sqrt{vx/U_0}}$$

and C_f can be given as a function of the Reynolds number:

$$C_f = \frac{\tau w}{\rho U_0^2/2} = 0.66 + Re_x^{-0.5}$$

If the boundary layer thickness δ is defined as the ordinate, where $U/U_0 = 0.99$ (a convectional definition), then

$$\eta = \frac{\delta}{\sqrt{vx/U_0}}$$

7.6.c *Boundary layers subjected to a pressure gradient (adverse or favorable)*

In this case, $U_0 \sim x^m$. Hartree has chosen a characteristic length scale that differs from the one introduced by Blasius:

$$\beta = \sqrt{\frac{2vx}{(m + 1)U_0}}$$

which gives $\beta = \sqrt{2vx/U_0}$, making reference to Blasius's case. The equation relative to F can now be written as

$$F''' - FF'' - \frac{2m}{m + 1} [(F')^2 - 1] = 0$$

and the velocity profiles for different values of m are presented in Fig. 7.9.

7.6.d *The role of surface forces in a boundary layer subjected to adverse pressure gradients*

Disregarding the problem of computations in connection with Hartree's equation, we examine particularly the role of the stress forces when an adverse pressure gradient is applied. This role is not trivial and quite interesting. Close to the wall, if we take into account the role of inertial and pressure forces, separation should occur immediately because the inertial forces are null at the wall. This prediction disagrees with both experimental data and computations. At a given station in a boundary layer, the pressure is constant and equal to the pressure measured in the inviscid flow just at the frontier ($\partial F/\partial y = 0$). The surface forces could reasonably be expected to play a favorable role in a more or less

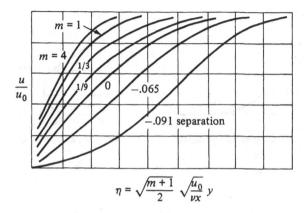

Fig. 7.9   Pressure gradient effects

extended zone near the wall. The action of the surface forces can be understood very quickly; a simple connection exists at the wall between the longitudinal pressure gradient and the transversal stress gradient, and therefore we can write

$$\frac{1}{\rho}\frac{\partial P}{\partial x} = \frac{\partial \tau}{\partial y} \qquad (7.26)$$

the inertial forces being null. Accordingly, the sign of $\partial \tau / \partial y$ is the same as the sign of $|\partial P/\partial x|_x$. At the wall the slope of the curve $\tau(y)$ is completely determined by the pressure distribution in the adjacent inviscid fluid. In as much as the stress force goes to zero at the external frontier of the boundary layer, two sorts of $\tau(y)$ distributions can be detected, as mentioned. They are sketched in Fig. 7.10a.

In Fig. 7.10b, a fluid element is subjected to two surface forces τ_A and τ_B. τ_B represents the action of the external region of the fluid; this force is oriented in the stream-flow direction. τ_A represents the action of the fluid located between the fluid element and the wall; this force acts in an opposite direction. Finally, the net balance of the surface forces applied to this fluid element is positive, which means that an entrainment force is acting on this part of fluid. Beyond the maximum, this net surface force becomes negative, but in this region inertial forces are important. The effects of pressure, inertial, and surface forces acting on a fluid element can lead to very different velocity profiles and must therefore be paid particular attention.

Even though this study is restricted to equilibrium states, it is interesting to examine the distribution of surface forces as we approach a separation point. The wall stress starts from a given value and decreases downstream. The transverse distribution of surface forces is organized

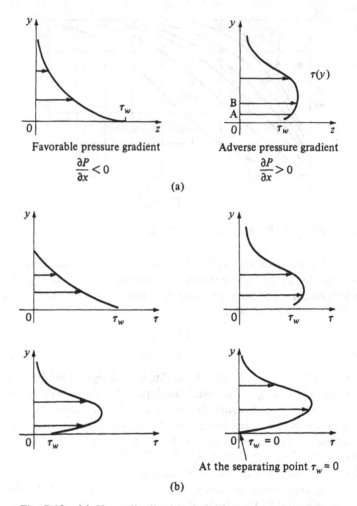

Fig. 7.10   (a) Shear distribution and (b) shear stress distribution in a boundary layer

in such a way as to have a favorable effect on the fluid streamlines next to the wall. Separation occurs at the location where the wall shear stress goes to zero (Fig. 7.10b).

We must pay special attention to the change of curvature in the velocity profiles corresponding to flows subjected to adverse pressure gradients. Due to a balance of surface forces, second derivatives appear in the Navier–Stokes equations that reinforce the role of the curvature of the velocity profile in this balance.

7.6.e   *Mathematical remarks*

The resolution of the Hartree (or Falker Skan) equations raises some difficulties, which we present briefly. Recall that a Cauchy problem assumes all the boundary conditions to be located at the same point. For an ordinary third-degree equation, we must know the first and the second derivative at the point under consideration. In the case of a boundary layer and for the function $F(\eta)$, two conditions must be verified at the wall, and the third condition corresponds to the external edge of the boundary layer.

$\eta = 0 \Rightarrow F(\eta) = 0$ first condition on the rate flow

$\eta = 0 \Rightarrow F'(\eta) = 0$ no-slip condition

$\eta = 1 \Rightarrow F'(\eta) = U_0$ condition at the external frontier

In other words, not all of the conditions relative to F are located at the same point. Such a situation could lead to an eigenvalue problem, but in that case no coefficient of the equation is adjustable and the problem could have no solution. In fact, it is possible to show that this problem does not introduce eigenvalue functions. We do not discuss this question in detail, but we note that the condition at the external frontier is not zero (the velocity does not go to zero as y tends to δ).* For this pattern of flow, we can say that this condition determines a sort of level regarding the solution (a level closely related to both the rate of flow and the momentum).

7.6.f   *Some comments on other patterns of flow*

Equilibrium profiles also exist for *mixing zones* (see figure). For this flow, the velocity goes from U_0 to zero. The pressure terms do not play any role in this development. An entrainment zone and the deceleration region are spreading in the downstream direction. The location of the mixing zone with respect to the trailing edge can be determined by the y component of the momentum equation.

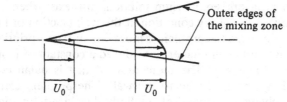

Outer edges of the mixing zone

U_0 U_0

---

* Where linear equations are concerned the eigenvalue problems have been examined thoroughly. For nonlinear equations, the approaches are delicate. The existence of a singular point is associated with these problems.

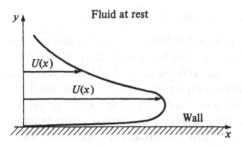

Fig. 7.11   Wall jet

As our previous remarks indicate, the problem of jets (see figure) should be more difficult because the velocity is null at the two edges of the jet. This symmetrical boundary condition should lead to an eigenvalue problem.

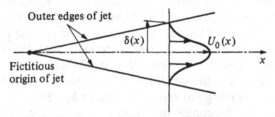

In fact, the amount of momentum is fixed by the conditions at the nozzle. Accordingly, the initial equations of motion are supplemented by this constant of motion, which is easily expressed when an equilibrium state exists:

$$\int_{-\infty}^{\infty} \beta U^2 \, dy = \beta U_0^2(x)\, \delta(x) \underline{\int_{-\infty}^{\infty} \frac{U^2}{U_0^2}\, d\!\left(\frac{y}{\delta}\right)} = \text{constant} \qquad (7.27)$$

Since the underlined integral is constant, the constant of motion becomes

$$\beta U_0^2(x)\, \delta(x) = \text{constant}$$

For boundary layers, equilibrium solutions also exist when a velocity is introduced as a boundary condition at the wall (suction or blowing). In the case of a *wall jet,* the problem is more difficult; no a priori constant of motion exists although Glauert introduced a constant of motion in a rather sophisticated way. The momentum deficit is balanced by the integral of drag forces applied to the wall. The problem can also be solved by applying mathematical methods developed for eigenvalue problems (Fig. 7.11).

If we assume that no momentum is introduced at the wall or at the outer edge, the equations of motion exhibit a singular point as y approaches infinity, introducing an eigenvalue problem whose detailed treatment is out of the range of this chapter.

Computations are carried out in the same way for the thermal problem as for the kinematic problem: We present only the essential features. The basic equations are supplemented by the equation of energy in a very simplified form:

$$U \frac{\partial U}{\partial x} + V \frac{\partial U}{\partial y} = -\frac{1}{\rho} \frac{\partial P}{\partial x} + \nu \frac{\partial^2 U}{\partial y^2} \tag{7.28}$$

$$\frac{\partial P}{\partial y} = 0 \tag{7.29}$$

$$U \frac{\partial T}{\partial x} + V \frac{\partial T}{\partial y} = \frac{\lambda}{\rho C} \frac{\partial^2 T}{\partial y^2} \tag{7.30}$$

$$\frac{\partial U}{\partial x} + \frac{\partial V}{\partial y} = 0 \tag{7.31}$$

in which λ is the thermal conductivity and C is the specific heat. Thermal effects are assumed to be weak, so the fluid density is constant and equal to ρ. An additional scaling has to be introduced. For temperature, it is convenient to take T_0 defined by

$$T_0 = T_w(x) - T_r$$

The wall temperature is denoted T_w and the temperature at the frontier is T_r, which is assumed to be constant. In this case, kinematic and thermal problems are uncoupled.

Taking adequate scaling into account, we introduce the function $g(\eta)$:

$$g(\eta) = \frac{T - T_r}{T_w - T_r}$$

with $\eta = y/\beta(x)$. If we set

$$\frac{\nu}{U_0 \beta \beta'} = a \qquad \frac{\beta U_0'}{\beta' U_0} = b \qquad \text{and} \qquad \frac{T_0'}{T_0} \frac{U_0 \beta^2}{\nu} = \text{constant}$$

we can write the energy equation as

$$g'' + Pr \frac{b + 1}{a} Fg' - Pr \, CF'g = 0 \tag{7.32}$$

With the Prandtl number defined as

$$Pr = \frac{\mu C_p}{\lambda} = \frac{\mu C}{\lambda}$$

and the velocity and temperature scales represented by the power laws

$$U_0 \sim x^m \qquad T_0 \sim x^n$$

the energy equation becomes

$$g'' + Pr \frac{m+1}{2} Fg' - Pr\, nF'g = 0 \qquad (7.33)$$

For a flat plate we deal with the simplified kinematic and thermal equations

$$2F''' + FF'' = 0 \qquad \hat{L}_c$$

$$2g'' + Pr\, Fg' = 0 \qquad \hat{L}_T$$

The continuity equation is taken into account by introducing a *pseudo stream function F*. The Prandtl number enables us to compare kinematic and thermal diffusion:

$$\frac{\mu/\rho}{\lambda/\rho C} = \frac{\mu C}{\lambda}$$

If the ratio is equal to unity, the two previous equations give

$$2F''' + FF'' = 0$$

$$2g'' + Fg' = 0$$

and the relationship between the two fields becomes evident:

$$g = F' \qquad \text{and} \qquad g = 1 - F'$$

The role of the Prandtl number will be examined later.

With air, whose Prandtl number is 0.7, the thermal boundary layer thickness is approximately the same as that of the kinematic boundary layer. However, we could list numerous cases in which the Prandtl number is very different from unity. For liquid metals, for example, the Prandtl number is very small; therefore, thermal diffusive effects dominate, and the thermal boundary layer grows rapidly in the flow field. Industrial applications can be considered with a view to increasing heat transfer. The Lewis number, whose role is similar to that of the Prandtl number, can be very different from unity in chemical reactors, and its role may be determinant for problems in flame stability.

7.7 COUPLED PHENOMENA

7.7.a *Kinematic boundary layer*

Viewed as a whole rather than in the light of a typical phenomenon, the boundary layer theory is significantly enlarged. Two mechanisms control the development of a boundary layer: an advective process associated with the undisturbed flow velocity U_0 (or any function of U_0) and a diffusion process that develops perpendicularly to the advective process.

In this section, we give other examples of coupled advective and diffusion processes, reexamining the boundary layer problem in the light of just this simple coupling. Therefore, we take a similar approach to laminar flames. It is easy to understand the method from these examples and to extrapolate to complex problems. This complexity is the consequence of intricacies relative to several physical phenomena acting in the same domain, the process intensities being strongly dependent on the different zones of this domain.

Operators corresponding to typical mechanisms must be identified whenever a complex phenomenon is examined. As these mechanisms appear in the same equation, their *characteristic times* must be of the same order of magnitude. In other cases, one of the phenomena dominates, and the equation degenerates into an equation of another kind. Bearing that in mind, we shall focus attention on the equation for the third component vorticity. Because the pressure terms do not appear, these developments also apply to boundary layers that are subjected to pressure gradients (adverse or favorable). The flow is assumed to be two dimensional:

$$\frac{\partial \zeta}{\partial t} + U \frac{\partial \zeta}{\partial x} + V \frac{\partial \zeta}{\partial y} = \nu \left[\frac{\partial^2 \zeta}{\partial x^2} + \frac{\partial^2 \zeta}{\partial y^2} \right] \qquad (7.34)$$

advective operator

diffusive operator

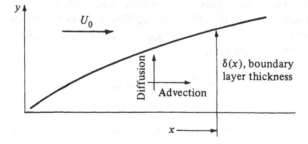

As usual, the boundary layer approximation assumes that advection predominates in the x direction and diffusion in the y direction (see figure). This *crossing mechanism* is typical of boundary layers encountered in fluid mechanics. The time scale of the two mechanisms can now be evaluated:

$$\tau_a \sim \frac{x}{U_0} \qquad \tau_d \sim \frac{\delta_T^2}{\alpha}$$

advective time   diffusion time

(The choice of U_0 could be discussed at a station, but we must remark that the velocity at any ordinate of this section is closely related to U_0.) These time scales characterize the two operators belonging to the vortex equation. They do not have to be of same order of magnitude. If they do not, the equation should degenerate into one that exhibits other mathematical and physical properties.

Finally, we obtain

$$\tau_d \sim \tau_a \Rightarrow \frac{\delta^2}{\nu} \sim \frac{x}{U_0} \Rightarrow \delta \sim \sqrt{\frac{\nu x}{U_0}}$$

The three parameters x, ν, and U_0 must be associated in this manner to maintain an equilibrium between the two mechanisms.

7.7.b *Thermal boundary layer*

These developments can easily be extended to a thermal boundary layer, whose equations have already been presented. If thermal effects are weak enough so that the kinematic field is not affected, the momentum and the continuity equations must be supplemented by the energy equation:

$$\frac{\partial T}{\partial t} + U\frac{\partial T}{\partial x} + V\frac{\partial T}{\partial y} = \frac{\lambda}{\rho C}\left[\frac{\partial^2 T}{\partial x^2} + \frac{\partial^2 T}{\partial y^2}\right] \qquad (7.35)$$

advective operator

diffusion operator

The same two mechanisms still compete, but diffusion is controlled by a coefficient $\alpha = \lambda/\rho C$. This new coefficient gives the thermal phenomenon its specificity. The time scales of the two mechanisms can be defined on dimensional grounds as

$$\tau_a \sim \frac{x}{U_0} \qquad \text{(advective time)}$$

$$\tau_d \sim \frac{\delta_T^2}{\alpha} \qquad \text{(diffusion time)}$$

Obviously these two scale times have to be of the same order of magnitude, which lead to

$$\tau_a \sim \tau_d \Rightarrow \frac{x}{U_0} \frac{\delta_T^2}{\alpha}$$

$$\delta_T \sim \sqrt{\frac{\alpha x}{U_0}}$$

The kinematic and the thermal thicknesses can now be compared. This results in

$$\frac{\delta_T}{\delta} = \sqrt{\frac{\lambda}{\rho C} \frac{x}{U_0} \frac{\rho}{\mu} \frac{U_0}{x}} = \sqrt{\frac{\lambda}{\mu C}} = \sqrt{\frac{1}{Pr}} \qquad (7.36)$$

Correspondingly, the Prandtl number, which is a measure of the ratio between the two diffusion coefficients ν and α, becomes

$$Pr = \frac{\nu}{\alpha} = \frac{\mu}{\rho} \frac{\rho C}{\lambda} = \frac{\mu c}{\lambda}$$

In the comparison between the two boundary layer thicknesses, the Prandtl number is the unique parameter, which appears because the adjective time is the same for the thermal and kinematic boundary layers.

The two boundary layer thicknesses are closely related because of the two starting equations, which are very similar to each other. The Prandtl number is equal to about 0.7 for air. It can be very small for liquid metals or very large for some oils. The relative position of the two boundary layers is presented in the figure.

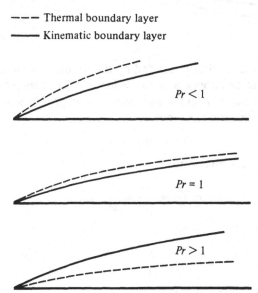

Any equation exhibiting the same structure can be treated in the same way. If a scalar quantity is conveyed by an advective field and if its diffusion process is controlled by a typical coefficient D, a new boundary layer δ_c takes place. Its thickness can be compared with the thickness of the kinematic boundary layer

$$\frac{\delta_c}{\delta} = \sqrt{\frac{1}{Le}}$$

in which the Lewis number (Le) plays a role similar to that of the Prandtl number. It controls the diffusion of chemical species.

7.7.c *Ekman layer in a rotating fluid*

Eckman layers, which develop in geophysical situations, are a typical example of layers for which rotation plays an essential role (see figure).

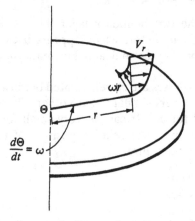

A simple case is the motion generated by the rotation of a disk. The radial and tangential components of the velocity V are represented at radius r. At some distance from the axis, the velocity profile is observed to be in an equilibrium state, so that

$$V_r = \omega r$$

If we introduce an advective time

$$\tau_a = \frac{r}{V_r} = \frac{r}{\omega r} = \frac{1}{\omega}$$

and a diffusive time

$$\tau_d \sim \frac{\delta^2}{\nu}$$

The comparison of τ_c and τ_d leads to

$$\delta \sim \sqrt{\frac{\nu}{\omega}}$$

7.7.d *Complex boundary layers*

To show the extent of the flexibility of the method we now apply it to materials exhibiting complex rheological properties. The stress at a point is considered to depend on the history of the strain, which is taken into account through the material derivatives of the strain tensor D, thus leading to the Rivlin tensors. With a material that exhibits a weak memory, it is possible to surmise that T depends both on D and on the first material derivative of D only. A rather cumbersome computation leads to an equation for two-dimensional flows. The material derivative, which is added to D in the constitutive law, leads to a partial differential equation of the third degree. A material derivative of the laplacian of ζ appears and is responsible for the change of degree in the equation. A new coefficient appears in connection with a weak memory effect; this fact can be interpreted as an elastic property of the fluid material. The behavior of a spring can be thought of as that of a system exhibiting a very strong memory, so the fluid under consideration behaves more or less like a spring.

We recall that the constitutive equation for a fluid is given by

$$T_{ij} = -P\delta_{ij} + 2\mu D_{ij} + \beta D_{ik}D_{kj} + \gamma \Delta_{ij}^{(2)} \tag{7.37}$$

where $A_{ij}^{(2)}$ is the second-order Rivlin–Erickson kinematic tensor, μ is the dynamic viscosity coefficient, and β and γ are additional phenomenological parameters.

In the case of a two-dimensional flow, we can write an equation for the ζ component of the vorticity:

$$\rho \underbrace{\frac{D\zeta}{Dt} = \mu\, \Delta\zeta}_{\substack{\text{all the terms exist in}\\ \text{the Navier–Stokes equations}}} + \underbrace{\gamma \frac{D}{Dt}[\Delta\zeta]}_{\substack{\text{this term displays the elastic}\\ \text{behavior of the fluid material}}} \tag{7.38}$$

$$\frac{\partial\zeta_i}{\partial t} + U_j \frac{\partial\zeta_i}{\partial x_j} = \nu\frac{\partial^2\zeta_i}{\partial x_j\,\partial x_j} + \frac{\nu\tau_0}{\rho}\left\{\frac{\partial}{\partial t} + U_j\frac{\partial}{\partial x_j}\right\}\frac{\partial^2\zeta_i}{\partial x_j\,\partial x_j} \tag{7.39}$$

Three characteristic times appear: an advective time $\tau_z \sim x/U_0$, a diffusion time $\tau_d \sim \delta_i^2/\nu$, and a scale time τ_e in related elastic effects, $\tau_e \sim U_0\delta_e^2/(\gamma/\rho)$. The Deborah number (De) is defined as the ratio of two characteristic times:

$$De = \frac{\tau_a}{\tau_d} = \frac{\tau_a\nu}{\delta_i^2}$$

De is null if no elastic effect exists in the fluid medium; in this case, a new boundary layer develops whose properties are different from those of a usual boundary layer (one that obeys Prandtl equations). This new

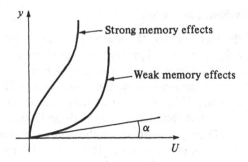

Fig. 7.12   Memory effects

phenomenon gives rise to an important question: What additional boundary conditions are needed to enable us to solve this third-order equation? We can try to predict the evolution of the velocity as the elastic properties of the fluid material become more and more predominant; but because the slope α of the velocity profile at the wall is closely related to memory effects, α should tend to $\pi/2$ as memory effects dominate (Fig. 7.12). On the other hand, we could also suspect the existence of equilibrium profiles if an additional parameter is introduced at the wall, for instance, by means of suction effects. The complexity of the rheological behavior leads to a rheological nonlinearity that should introduce an additional coupling process when turbulence takes place. In any case the problem is still open.

7.8 COMBUSTION PROBLEMS IN FLUID MECHANICS

Combustion problems are very difficult and are presently the aim of numerous investigations. They appear to be a particular case of the following problem: What is the influence of the kinematic properties of a medium on the development of chemical reactions? Such an investigation is based on the assumption that fundamental chemical problems are solved, but these problems deal only with reaction rate in an homogeneous medium at rest. These chemical properties are not discussed here, but let us just recall what the influence of the main parameter may be.

Disregarding the details of the chemical process, we surmise that the reaction rate strongly influences the geometry of the reaction zones. Flames are mainly characterized by high reaction rates and strong interactions between the chemical reaction and the kinematic properties of the medium. When two chemical species slowly interact, the reaction time is long and develops in an extended domain whose characteristic length L can be defined by

$$L = U\tau_r$$

in which U is the fluid velocity and τ_r is the reaction time. In that case, U is of the same order of magnitude as any other macroscale of the fluid motion. Briefly speaking, the chemical reaction develops in a volume. Moreover, in many chemical processes the role of the temperature on the reaction can be considered to be weak. As the reaction rate grows, the size of the domain becomes smaller and smaller. The initial volumes degenerate into a surface, or rather into a thin zone that can be regarded as a matching region that bridges the gap between an unburned-gas region and a burned-gas region. Upstream and downstream, the flame zone advective effects prevail whereas inside the flame zone advective and diffusive effects compete. The high reaction rate in the flame zone is kept high by heat release. All the mechanisms are strongly interconnected.

7.8.a   *Combustion in homogeneous mixtures*

In this case the gas mixture is assumed to be homogeneous. The experimenters agree on the general features of flame propagation, which can be summarized as follows: The reaction rate is a function of the mixture ratio and depends on temperature (Fig. 7.13). Flames propagate at measurable speed only if the temperature of the burned gas is quite high (greater than 1500 K for a hydrocarbon–air mixture). Consequently, for a given initial temperature, only a narrow range of mixture ratios on either side of the stochiometric state permits flame propagation (Fig. 7.14). The change of flame speed with composition is very rapid close to both limits.

7.8.b   *Combustion in a moving medium*

Attention is first focused on the energy balance. We use a one-dimensional analysis with a view to introducing an operating point. Accordingly, no reference is made to either laminar or turbulent flow. The heat release, which comes from the chemical reaction, depends on the temperature and can be represented by line 1 in Fig. 7.15.

For the temperature to increase, the gas crossing the flame zone requires an amount of heating that depends globally on flow rate. The lines 1 and 2 define an *operating point P*. If no interaction exists for increasing values of the flow rate, the flame is extinguished. An unstable zone is shown in the figure at P'.

The operating point corresponds to the case defined by the relation

$$q_m C(T_f - T_i) = \text{heat release}$$

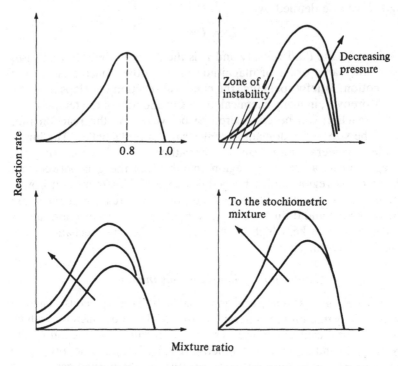

Fig. 7.13   Mixture-ratio effects, in which the change of the reaction rate is represented as a function of the mixture ratios

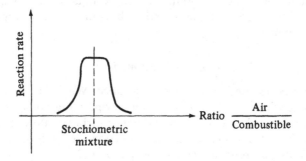

Fig. 7.14   Reaction rate as a function of mixture ratio

and the straight line corresponds to the amount of heating required by the flow rate crossing the flame:

$$q_m C(T - T_i) \qquad \text{(line 2)}$$

The stable point temperature is T_f, whereas for T_b no heat is released.

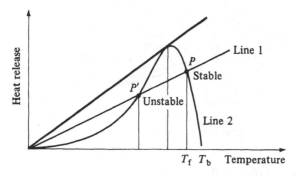

Fig. 7.15   Operating points

7.8.c   *Combustion processes*

We can separate combustion processes into four categories:

1. Homogeneous media, for which only the physical properties of the media play a role.
2. Flames resulting from a mixing process between a combustible gas and air. The flame mechanism is dominated by the mixing process, and the flame is located approximately in the zone where the stochiometric state is reached. These flames are termed *diffusion flames*.
3. If combustible gas and air are premixed, a steady flame corresponds to an equilibrium state, and diffusion and advective processes are in competition in the flame zone. The properties of these flames can be roughly analyzed and comparisons made with boundary layers. These flames are called *deflagration flames*.
4. Flames can result from an increase of temperature in connection with shock waves, and the speed of the flame is controlled by that of the shock wave. The combustion that is initiated by the shock is located just behind the shock. The flame lasts as long as is required to burn the combustile gas, that is, for a time closely related to the characteristic time of the chemical reaction.

Accordingly, the thickness of the flame zone is given by $\delta = \tau_r \times a$, where a is the celerity of the shock wave, which is at least of the same order of magnitude as sound speed (Fig. 7.16). In any case, this propagation is much faster than that of a deflagration. The phenomenon is called *detonation*. In combustion engines, spontaneous ignition of combustible gas before contact with the flame front may occur. This is the mechanism of *knock*. When this phenomenon occurs, violent pressure fluctuations occur. The tendency toward spontaneous ignition ahead of the flame is increased when the flame begins to travel across the combustion chamber, because the volume of this chamber hardly changes and so the pressure rises.

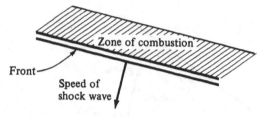

Fig. 7.16   Combustion behind a shock wave

7.8.d   *Diffusion flames*

Diffusion flames correspond to a combustible gas emerging from an opening and mingling with surrounding air. The mechanism taking place in the jet operates on the mixture, and surfaces of equal concentration can be determined (Fig. 7.17). One of these surfaces corresponds to the stochiometric state and plays a dominant role in combustion processes because the flame is necessarily located in a narrow zone on either side of this stochiometric surface. Due to small delay in ignition, this surface flame is located in the surrounding region corresponding to the stochiometric surface. The shape of diffusion flames can be approximately determined as a hot jet emerging in air at ambient temperature. In this case, the outer edge of the flow is characterized by a temperature gradient. If the Lewis number is not too different from the Prandtl number, a surface of equal temperature corresponds roughly to the stochiometric surface. In such diffusion flames, mechanical problems become predominant.

7.8.e   *Premixed flames*

Most industrial flames are diffusion flames. However, in some cases a double process is introduced. A premixed mixture whose chemical composition is too lean to burn is introduced, and the flame is established in a second mixing zone and can be considered as a diffusion flame. Stability is improved by this two-step method.

Premixed flames result from two mechanisms that are in strong competition in the flame zone: an advective process and a diffusion process. The chemical reaction also introduces a typical characteristic time by heat release in the same zone. A one-dimensional analysis of this sort of flame is sufficient to understand physically what kinds of relationships exist with boundary layer theory. First, let us examine the role of the temperature:

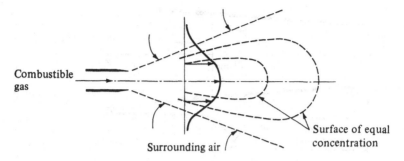

Combustible
gas

Surrounding air

Surface of equal
concentration

Fig. 7.17   Jet structure

$$U\frac{dT}{dx} = \frac{\lambda}{\rho C}\frac{d^2T}{dx^2} + Q \qquad (7.40)$$

$$= \alpha\frac{d^2T}{dx^2} + Q \qquad (7.41)$$

advection   diffusion   heat release

As mentioned, two orthogonal mechanisms are in competition in a
boundary layer: advection and diffusion. These mechanisms also compete in the flame zone (see Fig. 7.18), but in this case they are collinear.
Moreover, a strong heat release is associated with the reaction, which
generates a more or less extended preheated zone. An analysis of the
phenomenon is based on three characteristic times:

advective time $\sim \delta_l/U_l;\ U_l = U$
diffusive time $\sim \delta_l^2/\alpha$
chemical characteristic time, τ_r

All the phenomena, being coupled, are of the same order of magnitude,
which leads to

$$\frac{\delta_l}{U_l} \sim \frac{\delta_l^2}{\alpha} \sim \tau_r$$

which yields

$$\delta_l = \sqrt{\alpha\tau_r}$$

$$\delta_l \sim \frac{\alpha}{U_l} \sim \frac{\alpha}{U}$$

U is also the laminar burning velocity U_l (speed of flame) since the flame
is supposed to be steady.

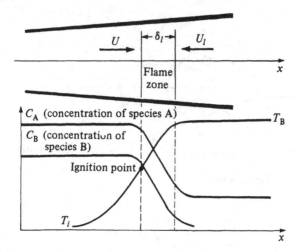

Fig. 7.18 Flame location

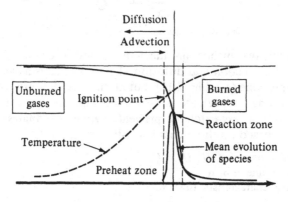

Fig. 7.19 Combustion zone scheme

A one-dimensional equation can also be written for the chemical species,

$$U \frac{dC_i}{dx} = D \frac{d^2C_i}{dx^2} + Q_{C_i}$$

in which a new diffusion coefficient D has been introduced. The ratio α/D is termed the Lewis number Le. Disregarding the details, we can demonstrate the importance of this dimensionless number by numerical analysis. The flame zone is essentially a meeting zone where the concentration of species and the temperatures must remain within a given range. For instance, a flame cannot exist if the temperature field is spreading due to high values of the conductivity of the medium. Low

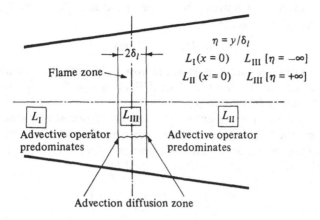

Fig. 7.20   Flame zone location

values of the Lewis number for the fuel can cause flame instabilities. In fact, the diffusion zone is more extended than the reactive zone, as shown in Fig. 7.19, and in the flame zone (Fig. 7.20) we can also introduce an expanded operator by using a scaling based on flame thickness.

7.8.f   *Flame Speed*

Measurements of flame speeds can be made either by using a divergent duct or by measuring the angle of the flame at the exit of a burner. In the first case, a steady phenomenon is observed; the speed of the flame is equal in magnitude to the flow speed but opposite in direction (see top part of figure).

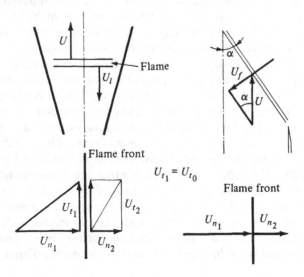

At the nozzle of a burner, the flame speed may be connected to α (see bottom part of figure):

$$U_l = U_{\text{flame}} = U \sin \alpha$$

Crossing an oblique flame, part of the flow is deflected, as occurs when the flow crosses a shock wave. This deflection is a direct result of the change of density from ρ_i to ρ_f.

Because of the heating expansion and acceleration of gases in the flame front, there is a pressure difference across the flame front. It can be shown that the pressure difference between that of the initial unburnt gas and that of the final burnt gas is equal to

$$P_i - P_f = \rho_i U_i^2 \left(\frac{U_f}{U_l} - 1 \right)$$

This relation must be complemented by the continuity equation

$$\rho_f \, U_f = \rho_i U_i$$

Taking into account that $U_i = U_l$ (the speed velocity U_i is equal to the flame velocity, but in a counterdirection so that the flame is steady), we find that

$$P_i - P_f = \rho_i U_i^2 \left[\frac{\rho_i}{\rho_f} - 1 \right]$$

which shows that the speed of the flame can be determined from pressure measurements.

This deflection of the flow by the flame front also interferes with stability (Fig. 7.21). The velocity of fluid particles is reduced in the cross-hatched zone (the flame front behaves as a blunt body in these regions), whereas this same velocity increases in other zones. Accordingly, the flame front should, respectively, move downstream and upstream, so the initial waving shape of the flame front should become more and more pronounced if other parameters do not come into play.

The role of the Prandtl number is currently emphasized in all problems of boundary layers, and its importance has been displayed in an intuitive way. The role of the Lewis number is also fundamental, and we can expect it to be strongly involved in all the problems of local stability of flames. If the flame front is locally distorted as indicated in Fig. 7.21, we suspect that the diffusion processes across the flame significantly increase due to the extension of the surface of the front. If the Lewis number is very different from unity, diffusion processes are modified in a selective way. Some species such as oxygen take a particular advantage of this opportunity. This diffusion coefficient of fuel being small, the relative concentration of fuel increases in the cross-hatched area, so the

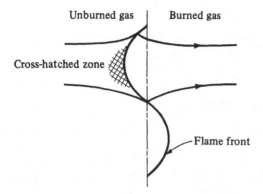

Fig. 7.21   Flame front

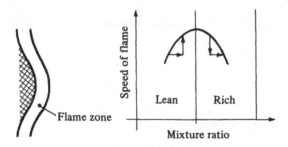

Fig. 7.22   Influence of mixture ratio

mixture becomes richer and richer. If the initial mixture is lean, the speed of the flame front increases locally, the waving motion is emphasized, and the system becomes unstable. If the mixture is initially rich, the speed of the flame decreases, and the initial disturbance tends to be removed (Fig. 7.22).

This simple example emphasizes the role of the Lewis number in complex diffusion processes. The influence of the richness of the mixture on the flame speed and a diffusion process acting in a selective way are the two leading factors of this phenomenon. The shape of the flame front also influences the local thickness of the preheat zone.

7.9   THREE-DIMENSIONAL BOUNDARY LAYERS

Although we do not intend to treat the problem of three-dimensional boundary layers, it seems useful to make some elementary remarks that

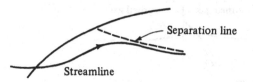

Fig. 7.23   Location of separation

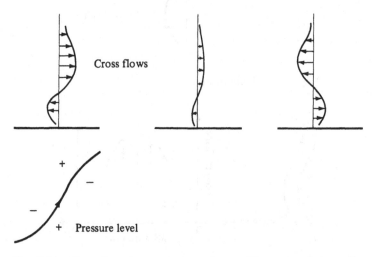

Fig. 7.24   Cross flow due to pressure forces. The external streamline
passes through an inflection front.

are of interest for industrial applications. The problem of separation
must be treated in a different way from two-dimensional flows because
the separation point appears to be a bifurcation point from which two
distinct flows emerge, and the evolution of the wall shear stress is not
directly related to the occurrence of a separation point (Fig. 7.23).

The existence of cross flows is a consequence of inertial effects. Ac-
cording to the terms of advective acceleration $U_j \, \partial U_i / \partial x_j$, inertial forces
depend on the velocity U_j, so these inertial forces are null at the wall
and are weak in a more or less extended area near the wall. Therefore,
the inner layer boundary responds quickly to pressure forces, and the
outer part of the boundary layer responds slowly to these pressure forces,
due to inertial forces. In other words, the time response of a boundary
layer that is subjected to an external disturbance depends on the local
inertial forces. This generates cross flows, as indicated in Fig. 7.24.

Spilling breaking waves. Waves are propagating from left to right. Wavelength = 0.75 m. Photograph by Ming-Yang Su. Reproduced by permission from M. van Dyke, *An Album of Fluid Motion.* Parabolic Press, Stanford, California, 1982, p. 114.

8   Wave Propagation

Much of the analysis of wave motion in acoustic and electromagnetic fields uses common physical concepts, and the mathematical nature of the equations is the same. They are partial differential equations of the hyperbolic type. Their characteristic lines can be identified as wave fronts, which lets us separate an undisturbed from a disturbed region; the disturbances emanating from the source can be detected in the disturbed region only.

Mathematical connections correspond to similar physical mechanisms. Wave propagation assumes reversible interactions between two kinds of energy. This reversibility exists when both electromagnetic and acoustic waves are involved, leading us briefly to examine the case of Alfvén waves, which result from an interaction between magnetic and inertial forces in fluid motions.

Wave propagation in an elastic medium is described in elementary treatises, which often use a spring to demonstrate the most essential features. A mechanical disturbance is generated at point A; two waves travel from A. By means of these two wave fronts we can separate the part of the spring that has been disturbed at time t from that which is not yet disturbed. Elastic waves are characterized by a velocity that depends both on the rigidity of the spring and on its mass (Fig. 8.1).

To remove secondary effects linked to the reflection of the waves, the spring is assumed to be unlimited in length. Potential and kinetic energy are closely associated, so the wave propagation is a straightforward consequence of exchanges between these two kinds of energy. In a continuous medium the complexity of the wave system depends on the ability of the medium to store several sorts of energy.

In this chapter, we consider wave propagation as a whole and focus on common features. Interactions between sound waves and turbulent media in which deterministic and random phenomena interact with each other are also outlined. The equations of motion are treated in a linear form, which requires a disturbance to be of small amplitude. Moreover, we assume the derivatives of small quantities to be of the same order of magnitude as the function itself. This assumption is somewhat questionable and unfortunately cannot be supported by mathematical de-

211

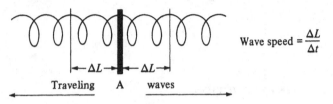

$$\text{Wave speed} = \frac{\Delta L}{\Delta t}$$

Fig. 8.1 Waves in a spring

velopments. We will consider it to be genuine here if theoretical findings are in good agreement with experimental data.

8.2 SOUND WAVES

We introduce the problem of sound wave propagation by making the following assumptions: The medium in which the sound waves travel is perfect, so we can neglect viscosity and thermal conductivity. The fluid is initially at rest and has pressure p_0, density ρ_0, and temperature T_0. The medium is subjected to a small disturbance characterized by

$$u_i'(\mathbf{x}, t) \qquad \rho'(\mathbf{x}, t) \qquad p'(\mathbf{x}, t) \qquad T'(\mathbf{x}, t)$$

The role of body forces is neglected. The disturbance is of an order of magnitude equal to ε, so the acceleration term is equal to

$$\frac{\partial u_i'}{\partial t} \sim \varepsilon \left(u_j' \frac{\partial u_i'}{\partial x_j} \sim \varepsilon^2 \right)$$

With these assumptions, the momentum equation becomes

$$\frac{du_i'}{dt} \sim \frac{\partial u_i'}{\partial t} = -\frac{1}{\rho}\frac{\partial p}{\partial x_i} = -\frac{1}{\rho_0 \left(1 + \dfrac{\rho'}{\rho_0}\right)}\frac{\partial p'}{\partial x_i} \sim \frac{1}{\rho_0}\frac{\partial p'}{\partial x_i} \qquad (8.1)$$

in which $p = p_0 + p'$ and $\rho = \rho_0 + \rho'$.

The continuity equation can also be simplified to yield

$$\frac{\partial \rho}{\partial t} - \frac{\partial(\rho u_j)}{\partial x_j} = \frac{\partial(\rho_0 + \rho')}{\partial t} + \frac{\partial}{\partial x_j}(\rho_0 + \rho')u_j' = 0 \qquad (8.2)$$

$$\frac{\partial \rho'}{\partial t} = -\rho_0 \frac{\partial u_i'}{\partial x_j} \qquad (8.3)$$

If a relation exists between p' and ρ', we can treat the problem with only two dependent variables: u_i' and ρ'. Neglecting viscous and conductive effects we find this relation to be

$$p = k\rho^\gamma$$

Finally, the two leading equations can be written as

$$\frac{\partial u_i'}{\partial t} = -\frac{1}{\rho_0}\frac{dp'}{d\rho'} \qquad \frac{\partial \rho'}{\partial x_i} = -a^2\frac{\partial(\rho'/\rho_0)}{\partial x_i}$$

The pressure p depends on x through ρ, and we set $dp'/d\rho' = a^2$.
Introducing the new function $\rho'/\rho_0 = \rho^*$, we obtain

$$\frac{\partial u_i'}{\partial t} = -a^2\frac{\partial \rho^*}{\partial x_i}$$

$$\frac{\partial \rho^*}{\partial t} = -\frac{\partial u_i'}{\partial x_i}$$

Using cross derivatives, we can eliminate one of the two variables. If
we choose u_i', for example, the resulting equation is

$$\frac{\partial^2 u_i'}{\partial t\,\partial x_i} = -a^2\frac{\partial^2 \rho^*}{\partial x_i\,\partial x_i} = \frac{\partial^2 \rho^*}{\partial t^2}$$

or

$$\left\{\frac{\partial}{\partial t^2} - a^2\frac{\partial^2}{\partial x_i\,\partial x_i}\right\}\rho^* = 0 \qquad (8.4)$$

Among all the functions $\rho^*(x, t)$, a selection is made through the
algebraic operator (Dalembertian):

$$\left\{\frac{\partial}{\partial t^2} - a^2\frac{\partial^2}{\partial x_i\,\partial x_i}\right\}$$

because $p = k\rho^\gamma$, we find that

$$a^2 = \frac{dp'}{d\rho'} = \frac{dp}{d\rho} = k\gamma\rho^{\gamma-1} = k\gamma\frac{\rho^\gamma}{\rho} = \gamma\frac{p}{\rho} = \gamma RT$$

$$a^2 = \gamma RT_0$$

An identical equation can be written for u_i':

$$\left\{\frac{\partial^2}{\partial t^2} - a^2\frac{\partial^2}{\partial x_i\,\partial x_i}\right\}u_i' = 0 \qquad (8.5)$$

Remarks:

1. These equations could be deduced from the equations of motion without reference to a relation between p and ρ. The momentum and the conservation equations must be supplemented by the energy equation and the equation of state. If we neglect thermal conductivity and viscous effects in the momentum and energy equations, we have a straightforward treatment that leads to

$$\frac{\partial^2 u_i'}{\partial t^2} = \gamma R T_0 \frac{\partial^2 u_i'}{\partial x_j \, \partial x_j} \qquad (8.6)$$

2. Now consider plane waves, which obey similar equations:

$$\frac{\partial^2 u'}{\partial t^2} = a^2 \frac{\partial^2 u'}{\partial x^2}$$

The general solution can be written as

$$u' = f(x \pm at) \qquad (8.7)$$

in which f is an arbitrary function of either $(x + at)$ or $(x - at)$. That yields

$$u'(x, t) = f_1(x - at) + f_2(x + at) \qquad (8.8)$$

The general solution of the equation consists of two sets of plane waves: The first one travels in the x direction and the second in the opposite direction. It is clear that the function $f_1(x - at)$ can be associated with small disturbances traveling in the x direction because we can write

$$f_1(x - at) = f_1\left[x + l - a\left(t + \frac{l}{a} \right) \right]$$

The disturbance located at point $x + l$ at time $(t + l/a)$ has the same magnitude as that existing at point x at time t. The small disturbance is traveling along the x axis without alteration. The celerity of the disturbance equals a, which is the speed of sound propagation (Fig. 8.2).

3. u' and ρ^* obey equations that have the same structure. The derivatives of u with respect to time are linked to the derivatives of ρ^* with respect to space, and conversely. In fact, this structure accounts for the exchanges between two sorts of energy. It is more illuminating to consider the coupled system $u' \cdot T'$, in this case kinetic and thermal energy.

   This possibility of exchange between two sorts of energy that do not have the same nature is remarkable. It is encountered

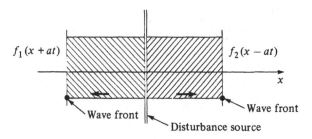

Fig. 8.2   General solution

in all the phenomena of wave propagation. This remark enables us to understand what the connections may be between phenomena that seem to differ.

8.3   SURFACE DISTURBANCES AND WAVE SPEED

We now write the equations of propagation of a small disturbance at the surface of a body of water, using the previous notation. The medium is assumed to be of constant density, and gravity forces predominate because exchanges between kinetic energy and gravity forces govern the propagation of these surface disturbances.

Let us consider a body of water of depth h in which a disturbance of height $h' = h'(x_1, x_2, t)$ above the mean (undisturbed) water level is propagated, and the x axis is taken along the bottom of the channel in which the waves are propagating in the direction of propagation. The wave motion obeys both the momentum and the continuity equation. We write

$$\frac{du_i'}{dt} = \frac{\partial u_i'}{\partial t} = -\frac{1}{\rho}\frac{\partial p}{\partial x_i} \qquad i \in 1, 2 \tag{8.9}$$

At point M, the pressure is equal to $p = \rho g h' + p_{\text{atm}}$, which assumes $\partial u_3'/\partial t$ to be small in comparison with g, yielding

$$\frac{\partial u_i'}{\partial t} = -g\frac{\partial h'}{\partial x_i}$$

The continuity equation can be written as

$$\frac{\partial u_i'}{\partial x_j} = 0$$

Because the role played by u_3' is typical, we write this equation as

$$u_3' = \frac{\partial h'}{\partial t} = -\int_0^h \frac{\partial u_j'}{\partial x_j}\,dx_3 \qquad j \in 1, 2$$

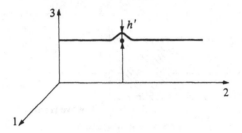

Fig. 8.3   Definition sketch for h'

and because $\partial u'_j / \partial x_j$ is independent of x_3, the preceding equation becomes

$$u'_3 = \frac{\partial h'}{\partial t} = -h\frac{\partial u'_j}{\partial x_j}$$

The motion obeys the following two equations (see Fig.8.3):

$$\frac{\partial u'_i}{\partial t} = -g\frac{\partial h'}{\partial x_i}$$

$$\frac{\partial h'}{\partial t} = -h\frac{\partial u'_j}{\partial x_j}$$

Using cross differentiation, we can remove one of the two variables.
Choosing, for instance, u'_i leads to

$$\frac{\partial^2 h'}{\partial t^2} = gh\frac{\partial^2 h'}{\partial x_j\,\partial x_j} \qquad j \in 1, 2$$

The role of the wave operator can still be emphasized

$$\left\{ \frac{\partial^2}{\partial t^2} - gh\frac{\partial^2}{\partial x_j\,\partial x_j} \right\} h' = 0 \tag{8.10}$$

The wave equation is homogeneous, and wave motion is generated by
an initial condition. No sources of disturbance exist at any time $t > t_0$.

To prepare for more sophisticated approaches, we propose a simple
introduction to characteristic curves.

Wave motion is assumed to be two dimensional, and the problem is
defined in the xy plane. The disturbance propagates in the x direction
(Fig. 8.4), obeying the following system of partial differential equations:

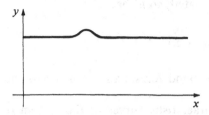

Fig. 8.4   Definition sketch

$$\frac{\partial h'}{\partial t} = -h \frac{\partial u'}{\partial x}$$

$$\frac{\partial u'}{\partial t} = -g \frac{\partial h'}{\partial x}$$

which is equivalent to the second-order equations

$$\frac{\partial^2 h'}{\partial t^2} = gh \frac{\partial^2 h'}{\partial x^2} \qquad \text{or} \qquad \frac{\partial^2 u'}{\partial t^2} = gh \frac{\partial^2 u'}{\partial x^2}$$

These equations correspond to wave propagation of celerity equal to $a = \sqrt{gh}$.

The general solution of the first equation can be written

$$h' = f_1(x - \sqrt{gh}\,t) + f_2(x + \sqrt{gh}\,t) \qquad (8.11)$$

The first set of waves travels in the positive x direction, and the second set in the opposite direction. The system of partial differential equations is verified if we write

$$u' = \sqrt{\frac{g}{h}}[f_1(x - \sqrt{gh}\,t) + f_2(x + \sqrt{gh}\,t)] \qquad (8.12)$$

It is easy to show that u' and h' are related at any point traveling with the speed of propagation of the disturbance.

If we examine the case of disturbance traveling in the x direction, the distance traveled as a function of time is $x = \sqrt{gh}\,t$. On this wave front, $x - \sqrt{gh}\,t = 0$ and $f_1(x - \sqrt{gh}\,t) = $ constant. Accordingly, we find that

$$h' = C_1 + f_2(x + \sqrt{gh}\,t) \Rightarrow f_2 = h' - C_1$$

$$u' = \sqrt{\frac{g}{h}}[C_1 + f_2(x + \sqrt{gh}\,t)]$$

The function f_2 can be eliminated, so u' becomes

$$u' = \sqrt{\frac{g}{h}}[-2C_1 + h']$$

A similar equation can be found for waves belonging to the second family.

Wave fronts are the characteristic curves of this system of partial differential equations. On a wave front, the physical properties of the medium u', h' are linked. This fundamental property will be demonstrated later. But it is informative to consider this sort of connection without sophisticated approaches. As with all gravity waves, the energy is divided between kinetic and potential parts. Reciprocal exchanges exist between these two forms, the two contributions being equal. Potential energy takes the place of thermal energy in the case of compressible flows.

The wave speed $a = \sqrt{gh}$ has a similar role to the speed of sound in compressible flows; thus, the speed of sound in one case or the speed of a surface disturbance in the other case can be used as references. If the fluid is in motion, two dimensionless parameters can be introduced: the Mach number (M) in the case of compressible flows,

$$M = \frac{|\mathbf{V}|}{a}$$

and the Froude number (Fr) for flows with a free surface,

$$Fr = \frac{|\mathbf{V}|}{\sqrt{gh}}$$

Wave motions become significant if we have

$$
\begin{array}{lll}
M \simeq 1 & \text{or} & M > 1 \quad \left\{ \begin{array}{l} \text{(transonic)} \\ \text{(supersonic flows)} \end{array} \right. \\
Fr \simeq 1 & \text{or} & Fr > 1 \quad \left\{ \begin{array}{l} \text{critical} \\ \text{supercritical flows} \end{array} \right.
\end{array}
$$

They are negligible if $M \ll 1$ (subsonic flows). If a body moves at supersonic speed in a medium at rest, it is possible to distinguish a disturbed region from an undisturbed one (see Fig. 8.5).

8.4   INTERNAL WAVES

Buoyancy forces, which arise as a result of density variation in a fluid subjected to gravity, produce a wide range of phenomena of importance in many branches of industry. Progress in this field has been made

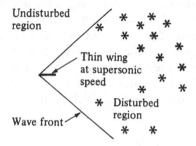

Fig. 8.5   Wave front

through the desire to solve practical problems arising in meteorology and hydraulic engineering.

This subject is closely related to the previous section because the problem consists of an interaction between kinetic energy and potential energy due to gravity; but in geophysical problems, this sort of phenomena interacts with Coriolis effects, so a global approach becomes very complex. If we limit the problem to energy exchanges between kinetic and gravitational fields, this subject can be considered an extension of the section on surface disturbances and wave speed. In that case, we examined the motion of an interface, but the influence of the upper medium on the free surface was neglected. Internal waves are a straightforward consequence of density gradients.

Natural bodies of fluid such as the atmosphere, oceans, and lakes are stably stratified; that is, their mean density decreases with height in most regions and most of the time. When they are disturbed in any way, internal waves are generated. A first approach to this phenomenon can be made using linear theories.

Many elementary properties of wave motions in stratified fluids can be introduced by considering waves at an interface between two superimposed layers. These waves are analogous to waves on a free water surface, and accordingly they can be seen as an extension of the previous concept of surface disturbances. In fact, the geometry of a surface at rest completely determines the direction of propagation of the waves on a free water surface. We emphasize that they are not the most general waves that can occur in a continuously stratified field. Energy can propagate through such a fluid at an angle to the horizontal, not just along the surface of constant density. To a large extent our intuition based on surface waves can be misleading, even through the mechanism of energy exchanges (potential and kinetic energy) is still acting.

Before dealing with the concept of internal waves, we must define the equilibrium state of the medium. The only external force field is gravity, which exerts a body force $\rho g(\rho g \delta_{i3})$ per unit volume on each element of fluid; ρ stands for the local density and **g** is the acceleration due to gravity. The effects to be described result from variations of ρ from point to point in the fluid. The nature of the fluid is not very important. Most of the ideas can be applied to liquids, in which density variations are due to differences of temperature and concentration, and to gases, in which there may be differences of composition or species.

We assume the fluid to be incompressible, which means that

$$\text{div } \mathbf{U} = 0 \qquad \frac{\partial u_j}{\partial x_j} = 0$$

and

$$\frac{D\rho}{Dt} = 0 \qquad \frac{\partial \rho}{\partial t} + u_j \frac{\partial \rho}{\partial x_j} = 0$$

where D/Dt denotes material derivatives. This results from the continuity equation in the form

$$\frac{\partial \rho}{\partial t} + u_j \frac{\partial \rho}{\partial x_j} + \rho \frac{\partial u_j}{\partial x_j} + 0 \tag{8.13}$$

which translates the fact that a fluid particle retains its density during its travel even though the density may not be constant at any point.

The assumption of incompressibility could appear to be drastic; but in fact, the compressibility of gases is significant only in deep layers, so for many purposes gases can be considered incompressible. Removing compressible modes, which introduce fluctuations that are colinear to the wave number vector κ, significantly reduces the difficulties encountered in treating the problem of turbulence in the atmosphere.

For a standard atmosphere, equilibrium states are defined which enable us to introduce departures from them. These problems obey the continuity equation in the form

$$\frac{\partial \rho}{\partial t} + u_j \frac{\partial \rho}{\partial x_j} = 0 \tag{8.14}$$

and the momentum equation

$$\rho \left[\frac{\partial u_i}{\partial t} + u_j \frac{\partial u_i}{\partial x_j} \right] = -\frac{\partial p}{\partial x_i} - \rho g \delta_{i3} \tag{8.15}$$

The x and y axes are in the horizontal plane, and z is vertical; p and ρ are now expanded about the values p_0 and ρ_0 that correspond to a reference state of hydrostatic equilibrium, for which

$$\frac{\partial p_0}{\partial z} = -\rho_0 g$$

Departure from the standard state is defined by the two variables p' and ρ'. The fluid is initially at rest $\mathbf{U} = \mathbf{u}'$, and so

$$p = p_0 + p' \qquad \rho = \rho_0 + \rho'$$

$$\mathbf{U} = \mathbf{u}'$$

The momentum equation becomes

$$\rho\left[\frac{\partial u_i'}{\partial t} + u_j'\frac{\partial u_i'}{\partial x_j}\right] = -\frac{\partial p'}{\partial x_i} - \rho'g\delta_{i3} \qquad \cdot(8.16)$$

and after linearization we have

$$\left[1 + \frac{\rho}{\rho_0}\right]\frac{\partial u_i'}{\partial t} = -\frac{1}{\rho_0}\frac{\partial p'}{\partial x_i} - \frac{\rho'}{\rho_0}g\,\delta_{i3}$$

When ρ'/ρ_0 is small, it produces only a small correction to the inertia term compared with a fluid of density ρ_0; but ρ' is a leading quantity in the buoyancy term. Boussinesq's approximation consists essentially of neglecting variations of density in so far as they affect inertia but retaining them in the buoyancy terms.

Finally, the linearized equations can be written

$$\frac{\partial u_i'}{\partial t} = -\frac{1}{\rho_0}\frac{\partial p'}{\partial x_i} - \frac{\rho'}{\rho_0}g\,\delta_{i3} \qquad (8.17)$$

$$\frac{\partial p'}{\partial t} + w\frac{\partial\rho_0}{\partial z} = 0 \qquad (8.18)$$

The changes of density at a point are due to bodily displacement of the mean density structure. Some basic parameters of heterogeneous flows can be put forward in an elementary way.

Now let us consider the motion of an element of inviscid fluid displaced vertically a small distance η from its equilibrium position in a stable environment. The small pressure fluctuations are also neglected. After rearrangements, the vertical component of the momentum equation, combined with the continuity equation, obeys the equation

$$\frac{\partial^2\eta}{\partial t^2} = -\frac{g}{\rho_0}\frac{\partial\rho_0}{\partial z}\eta \qquad (8.19)$$

The element oscillates in a simple periodic motion with an angular frequency given by

$$N = -\left[\frac{g}{\rho_0}\left|\frac{\partial\rho_0}{\partial z}\right|\right]^{1/2} \qquad (8.20)$$

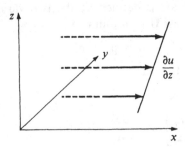

Fig. 8.6   Velocity gradient and Richardson number

where the absolute value is taken because $\partial \rho / \partial z < 0$ for stable stratification.

This buoyancy number is associated with the names of Brunt in meteorology and Väisälä in oceanography. The corresponding periods $2\pi/N$ are a few minutes in the atmosphere and many hours in the deep ocean.

In a shear flow the velocity gradient has the dimension of frequency, so we can introduce a dimensionless quantity, the Richardson number, Ri (Fig. 8.6).

$$Ri = \frac{N^2}{(\partial u/\partial z)^2} = -g \frac{\partial \rho/\partial z}{\rho(\partial u/\partial z)^2}$$

Other related numbers can be introduced, such as the flux Richardson number, which is the ratio of the removal rate of energy by buoyancy forces to the corresponding production by the shear.

Having outlined the overall problem, we now present some results related to waves at a boundary between homogeneous layers whose densities are ρ_0 and ρ_1 respectively, so that

$$\rho' = \rho_1 - \rho_0$$

The flow is assumed to be two dimensional with motion confined to the xz plane (z positive upward) and with crests consisting of parallel ridges running in the y direction. The linearized equations can be applied to each of the layers separately. Within the layers the motion is irrotational. Progressive waves in deep water can also be treated. A matching process is introduced at the separation zone. The displacement η is the same in the two layers; moreover, the pressure must be continuous at the interface. The phase velocity a^2 can be inferred:

$$a^2 = \frac{g\lambda}{2\pi} \frac{\rho_1 - \rho_0}{\rho_1 + \rho_0} = \frac{g\lambda}{2\pi} \frac{\rho'}{\rho_1 + \rho_0}$$

in which λ is the wavelength.

Such waves are called *dispersive* because their speeds depend on their wavelength. That leads us to distinguish between phase velocity and group velocity. Let us next consider two harmonic wave motions with the same amplitude A.

$$z_1 = A \cos(k_1 x - \omega_1 t)$$

$$z_2 = A \cos(k_2 x - \omega_2 t)$$

We set

$$k_1 - k_2 = \Delta k \qquad \frac{k_1 + k_2}{2} \neq k_1 \neq k_2$$

and

$$\omega_1 - \omega_2 = \Delta \omega \qquad \frac{\omega_1 + \omega_2}{2} \neq \omega_1 \neq \omega_2$$

so we have

$$z_1 + z_2 = 2A \cos \underbrace{\left[\frac{\Delta k}{2} x - \frac{\Delta \omega}{2} t \right]}_{\text{first term}} \cos \underbrace{\left[\frac{k_1 + k_2}{2} x - \frac{\omega_1 + \omega_2}{2} t \right]}_{\text{second term}} \qquad (8.21)$$

Due to the first term, we can globally observe an apparent motion with a very reduced speed that is called the group velocity a_g:

$$a_g = \frac{\Delta \omega}{\Delta k} \to \frac{d\omega}{dk}$$

The phase velocity is still given by

$$a = \frac{\omega}{k}$$

with $\omega = ka = 2\pi a/\lambda$ since $k = 2\pi/\lambda \to dk/k = -d\lambda/\lambda$. The group velocity a_g is equal to

$$a_g = \frac{d\omega}{dk} = a + k \frac{da}{dk}$$

$$= a - \lambda \frac{da}{d\lambda}$$

For the case under consideration,

$$a = \frac{g\lambda}{2\pi} \qquad (8.22)$$

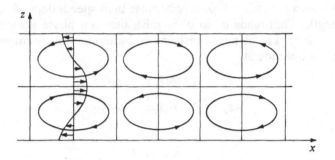

Fig. 8.7   Waves in a stratified fluid

we find that $a_g = \frac{1}{2}a$, a relation that also holds for deep-water surface waves. Energy is convected along the interface with the group velocity.

The small group of waves whose wave speed is a is found, in fact, at a distance of about $a_g t$ from the initial disturbance. For waves in a continuously stratified incompressible fluid, an associated pattern of displacement is shown in Fig. 8.7. A cellular standing wave is represented by horizontal and vertical wavelengths.

Various finite amplitude phenomena exist in a stratified fluid, most of which occur in the context of hydraulic engineering: for example, in the prediction of the velocity of intrusion of saline water into a lock filled with fresh water. One layer can be withdrawn from a stratified fluid without removing an adjacent layer.

8.5   PRESSURE WAVE IN AN ELASTIC DUCT

Pressure waves in elastic ducts are interesting in the contexts of both hydraulic engineering and biological sciences. Any change in flow rate generates a pressure wave in the duct. The intensity of this pressure wave largely depends on the elasticity of the duct.

The velocity disturbance u' is associated with the pressure fluctuation p' through the Euler equation in its linearized form:

$$\rho_0 \frac{\partial u'}{\partial t} + \frac{\partial p'}{\partial x} = 0 \qquad (8.23)$$

Disturbances are measured with respect to an initial state, so Eq. 8.23 becomes

$$p = p_0 + p' \qquad \rho = \rho_0 + \rho'$$

The continuity equation can be written as

$$\rho u' S \, dt - \left[\rho u' S + \frac{\partial}{\partial x}(\rho u' S) \, dx \right] dt = -\frac{\partial}{\partial t}(\rho S \, dx) \, dt$$

which finally becomes

$$\frac{\partial}{\partial t}(\rho S) + \frac{\partial}{\partial x}(\rho u' S) = 0 \qquad (8.24)$$

For liquids, a simple relation exists between ρ' and p':

$$\rho' = \chi \rho_0 p'$$

Assuming the duct to be elastic, we write

$$\frac{S'}{S_0} = \frac{1}{E} \frac{D}{e} p'$$

in which D is the diameter of the duct, e the thickness of the wall; and E the elastic modulus of the duct material. We find that

$$\rho S = (\rho_0 + \rho')(S_0 + s') \approx \rho_0 S_0 + \rho S' + S_0 \rho'$$

$$\approx \rho_0 S_0 \left[1 + \left(\chi + \frac{1}{E} \frac{D}{e} \right) p' \right]$$

The continuity equation in its modified form is

$$\left(\chi + \frac{1}{E} \frac{D}{e} \right) \frac{\partial p'}{\partial t} + \frac{\partial u'}{\partial x} = 0$$

which must be associated with the momentum equation

$$\rho_0 \frac{\partial u'}{\partial t} + \frac{\partial p'}{\partial x} = 0$$

This leads to a wave equation for each component of the disturbance, and

$$\left\{ \rho_0 \left(\chi + \frac{1}{E} \frac{D}{e} \right) \right\} \frac{\partial^2 p'}{\partial t^2} = \frac{\partial^2 p'}{\partial x^2}$$

The speed of the pressure wave is equal to

$$a = \sqrt{\frac{1}{\rho_0 \left(\chi + \frac{1}{E} \frac{D}{e} \right)}}$$

The two variables p' and u' are linked on a wave front:

$$\frac{p'}{\rho_0 g} = \frac{au'}{g}$$

The pressure disturbance is linearly linked to u' through the wave speed a, which depends on the modulus elasticity of the duct. The speed of the wave is significantly altered by the straining of the duct. With a rigid duct, $a \simeq 1400$ m/s; in practice this value does not exceed 1000 m/s in a penstock.

8.6 ALFVÉN WAVES

Magnetohydrodynamics, a subject that has undergone an explosive development in the last few decades, is the branch of continuum mechanics that deals with the motion of an electrically conducting fluid in the presence of a magnetic field. The motion of conducting material across the magnetic lines of force creates potential differences that, in general, cause electric currents to flow. The magnetic fields associated with these currents modify in turn the magnetic field that creates them. In other words, the fluid flow alters the electromagnetic state of the system. On the other hand, the flow of electric current across a magnetic field is associated with a body force, the so-called Lorentz force, that influences the fluid flow. It is this interdependence of hydrodynamics and electrodynamics that defines and characterizes magnetohydrodynamics.

Where wave propagations are concerned, common features between hydrodynamics and electrodynamics can be detected, especially the connection between two sorts of energy that interact. More possibilities exist for storing energy when electric and magnetic fields interact because in them other kinds of interactions also exist, such as Alfvén waves, which we now examine.

8.6.a *Equations of electromagnetism (Maxwell equations)*

For a medium at rest, the Maxwell equations governing an electromagnetic field are

$$\int_C \mathbf{E} \, \mathbf{dl} = \iint_S \hat{\mu} \, \frac{\partial \mathbf{H}}{\partial t} \, \mathbf{dS} \Rightarrow \text{curl } \mathbf{E} = -\mu \, \frac{\partial \mathbf{H}}{\partial t}$$

$$\iint_S \frac{\partial \mathbf{E}}{\partial t} \, \mathbf{dS} = \int_C \mathbf{H} \, \mathbf{dl} \Rightarrow \varepsilon \, \frac{\partial \mathbf{E}}{\partial t} = \text{curl } \mathbf{H}$$

The last equation corresponds to the concept of displacement currents, which are assumed to have the same properties as electric currents.

Significant displacement currents are always associated with a very rapid evolution of the electrical field; for a long time that made their existence questionable.

If we assume that the detection of electromagnetic waves confirms the existence of displacement currents, we can associate the temporal evolution of **E** with a spatial evolution of **H**, which completes the first law, in which a temporal evolution of **H** is associated with a spatial evolution of **E**. In fact, these two equations lead to wave propagation.

First of all we assume the medium to be homogeneous and isotropic. In reference to continuum media, $\hat{\mu}$ and ε, which denote magnetic permeability and electric constant, respectively, are independent of x; moreover, $\hat{\mu}$ and ε are scalar quantities. This leads to **B** = $\hat{\mu}$**H** and **D** = ε**E**. Eliminating the magnetic field from the previous equations yields

$$\left[\frac{\partial^2}{\partial t^2} - \frac{1}{\varepsilon\hat{\mu}} \frac{\partial^2}{\partial x_j\, \partial x_j} \right] \mathbf{E} = \mathbf{grad}\ \mathrm{div}\ \mathbf{E}$$

If no isolated charge exists in the medium we have div **E** = 0, and

$$\left[\frac{\partial^2}{\partial t^2} - \frac{1}{\varepsilon\hat{\mu}} \frac{\partial^2}{\partial x_j\, \partial x_j} \right] \mathbf{E} = 0 \qquad (8.25)$$

which is the equation for electromagnetic waves.

8.6.b *Interactions between mechanical and electrical effects*

A simple example serves as guidance for interactions between mechanical and electrical effects. A solid conducting disk D is mounted on an axle AA' that is rotated by the application of a constant torque. The velocity U is in the direction indicated in Fig. 8.8. A magnetic field $\mathbf{B}_0$, parallel to the axle, cuts the plane of D everywhere, as shown. A potential gradient $\mathbf{U} \times \mathbf{B}_0$ is created that is directed toward the periphery of D. Positive charges are therefore accumulated in the periphery of the disk and negative charges in the axle region. If we complete the electric circuit by connecting the sliding brushes with a wire as in the figure, an electric current will flow through the circuit. This current produces a contribution **b** to the magnetic field **B**; that is, the field cutting the disk D is no longer $\mathbf{B}_0$ but is $\mathbf{B} = \mathbf{B}_0 + \mathbf{b}$. Thus, the motion characterized by U, alters the initial magnetic field **B**.

Now consider the effect of the field **B** on the motion. The fact that a current of density **j** conveyed by the disk crosses the global field **B** implies that a Lorentz force $\mathbf{j} \times \mathbf{B}$ per unit volume acts on the disk. This force is directed in opposition to U, and in a steady state its integrated moment

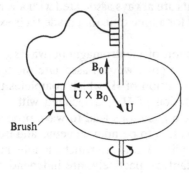

Fig. 8.8   Contribution to original magnetic field

about the axle AA' must balance the applied torque exactly. In a fluid medium, Lorentz forces applied to particles can significantly modify their trajectories, which makes the problem even more complex.

8.6.c   *The equations of electromagnetism in a moving medium*

The velocity of the medium is still termed **U**. We assume the electric current density **j** to be related to **E** by the relation **j** = σ**E** (Ohm's law in an isotropic medium), where σ is the electrical conductivity and its inverse σ^{-1} is the resistivity. Finally, $\eta = 1/\sigma\hat{\mu}$ is the magnetic diffusivity.

Most magnetohydrodynamics applications use a *quasi-steady approximation*, which assumes that no significant alteration of the system occurs in the time taken by the light to cross it. This approximation filters out electromagnetic radiation, which we can consider to be instantaneously propagated. The displacement current is omitted, and field **B** is instantaneously established. The consequences of this hypothesis follow.

According to standard electromagnetic theory, we have

$$\mathbf{E}' = (1 - \gamma_u) \frac{\mathbf{U} \cdot \mathbf{E}}{\mathbf{U} \cdot \mathbf{U}} \mathbf{U} + \gamma_u (\mathbf{E} + \mathbf{U} \times \mathbf{B})$$

$$\mathbf{B}' = (1 - \gamma_u) \frac{\mathbf{U} \cdot \mathbf{B}}{\mathbf{U} \cdot \mathbf{U}} \mathbf{U} + \gamma_u \left(\mathbf{B} - \frac{\mathbf{U} \times \mathbf{B}}{c^2} \right)$$

where

$$\gamma_u = \left(1 - \frac{\mathbf{U} \cdot \mathbf{U}}{c^2} \right)^{-1/2}$$

and c denotes the speed of light in a vacuum. With moderate values of $|U|$ only, these equations become

$$\mathbf{E'} = \mathbf{E} + \mathbf{U} \times \mathbf{B} \qquad \mathbf{B} = \mathbf{B'}$$

$$\text{curl } \mathbf{B'} = \text{curl } \mathbf{B}$$

and because $\mathbf{j'} = \text{curl } \mathbf{B'}$ and $\mathbf{j'} = \text{curl } \mathbf{B}$, we have $\mathbf{j} = \mathbf{j'}$.

Note that Ohm's law can be applied only in a frame of reference that moves locally with the fluid at the point of interest:

$$\mathbf{j'} = \mathbf{j} = \sigma\mathbf{E'} = \sigma(\mathbf{E} + \mathbf{U} \times \mathbf{B})$$

The field $\mathbf{U} \times \mathbf{B}$ is sometimes called the *induced electric field,* and the field $\mathbf{E'}$ is the *effective electric field.* Finally, $\mathbf{E'}$ is the electric field that induces the current flow in the moving fluid. All the quantities in the last equation are determined in the laboratory.

These equations are, of course, supplemented by the first Maxwell equation

$$\text{curl } \mathbf{E} = -\hat{\mu}\,\frac{\partial \mathbf{H}}{\partial t}$$

which can be considered a consequence of inertial properties of the magnetic field.

The equations for determining the fluid motion can be written as

$$\frac{\partial \rho}{\partial t} + \text{div}(\rho\mathbf{U}) = 0 \tag{8.26}$$

$$\rho\,\frac{d\mathbf{U}}{dt} = -\,\text{grad }p + \rho\mathbf{g} + \mathbf{f} + \hat{\mu}\,\mathbf{j} \times \mathbf{H} \tag{8.27}$$

Finally, three unknown quantities $\mathbf{E}$, $\mathbf{H}$, and $\mathbf{j}$ come into play, so we introduce three additional equations:

$$\text{curl } \mathbf{E} = -\mu\,\frac{\partial \mathbf{H}}{\partial t}$$

$$\hat{\mu}\,\text{curl } \mathbf{H} = 4\pi\mathbf{j}$$

$$\mathbf{j} = \sigma(\mathbf{E} + \mathbf{U} \times \mathbf{B})$$

8.6.d *Consequences of these equations on the electromagnetic field*

Let us suppose the space to be a homogeneous solid conductor so that σ becomes a constant. For the three equations,

$$\textbf{curl H} = \textbf{j} \qquad \text{div } \textbf{j} = 0$$

$$\textbf{curl E} = -\hat{\mu} \, \frac{\partial \textbf{H}}{\partial t} \qquad \text{div } \textbf{H} = 0$$

$$\textbf{j} = \sigma(\textbf{E} + \hat{\mu}\textbf{U} \times \textbf{H}) \tag{8.28}$$

we obtain

$$\frac{\partial \textbf{H}}{\partial t} = \textbf{curl}\left(\textbf{U} \times \textbf{H} - \frac{\textbf{j}}{\sigma\hat{\mu}}\right)$$

$$= \textbf{curl}(\textbf{U} \times \textbf{H}) + \eta \, \nabla^2 \textbf{H}$$

with

$$\eta = (\hat{\mu}\sigma)^{-1}$$

This equation gives the temporal evolution of the magnetic field. The mechanism reveals a competition between the entrainment effect by the fluid and diffusive effects. If we introduce a length scale L in connection with the size of the fluid, we can compare these two kinds of effects by means of a magnetic Reynolds number, which plays the same role as the Reynolds fluid number introduced in fluid mechanics:

$$Re_\eta = \frac{L|\textbf{U}|}{\eta}$$

If entrainment effects are very small, the previous equation becomes

$$\frac{\partial H}{\partial t} = \eta \, \nabla^2 H$$

A characteristic time can be introduced in connection with diffusion effects:

$$\tau_{\text{d}} \sim \frac{L^2}{\eta}$$

Another limiting factor corresponds to the case in which the medium in motion exhibits a high conductivity σ and so η can be considered small. The entrainment of the magnetic field by the medium in motion becomes predominant, and the equation is reduced to

$$\frac{\partial \textbf{H}}{\partial t} = \textbf{curl}(\textbf{U} \times \textbf{H})$$

After some rearrangements that introduce the continuity equation and div $\textbf{H} = 0$, we find that

$$\frac{d}{dt}\frac{\textbf{H}}{\rho} = \left(\frac{\textbf{H}}{\rho} \cdot \textbf{grad}\right)\textbf{U}$$

A characteristic time can be introduced in connection with entrainment effects:

$$\tau_{ad} \sim \frac{L}{|U|}$$

This leads to a situation very similar to that described in boundary layer theory. Magnetic fields that are completely driven by motion are said to be "frozen" with respect to the fluid particles. The arguments used for boundary layer theory still apply here.

8.6.e *Magnetohydrodynamic waves*

The previous developments introduced the reader to Alfvén waves. Alfvén explains the existence of waves as a consequence of interactions between the inertial properties of the magnetic field and that of the fluid. Let us start from the equations that represent these inertial properties:

$$\frac{\partial \mathbf{H}}{\partial t} = \mathbf{curl}(\mathbf{U} \times \mathbf{H}) \qquad (8.29)$$

and

$$\rho \frac{d\mathbf{U}}{dt} = -\mathbf{grad}\, p + \rho \mathbf{g} + \hat{\mu}(\mathbf{curl}\, \mathbf{H}) \times \mathbf{H} \qquad (8.30)$$

Neglecting the dissipative effects, we assume an extended mass of liquid to be immersed in a uniform magnetic field $\mathbf{H}_0$. We introduce two small disturbances into a part of the fluid medium: a velocity perturbation $\mathbf{u}'$ and a small magnetic disturbance $\mathbf{h}'$. These disturbances are related by the two equations

$$\frac{\partial \mathbf{h}'}{\partial t} = \mathbf{curl}[\mathbf{u}' \times (\mathbf{H}^0 + \mathbf{h}')]$$

$$\rho \frac{d\mathbf{u}'}{dt} = -\mathbf{grad}\, P + \rho \mathbf{g} + \hat{\mu}(\mathbf{curl}\, \mathbf{h}') \times (\mathbf{H}^0 + \mathbf{h}') \qquad (8.31)$$

A temporal evolution of $\mathbf{h}'$ is connected with a spacial distribution of $\mathbf{u}'$ conversely. In this computation we are interested only in wave motions that result from this coupling system, which means that the fluid is assumed to be incompressible div $\mathbf{U} = 0$ and at rest. Taking account also of div $\mathbf{h}' = 0$, we have

$$\frac{\partial \mathbf{h}'}{\partial t} = (\mathbf{H}^0 \cdot \mathbf{grad})\mathbf{U}$$

$$\rho \, \frac{\partial \mathbf{u}'}{\partial t} = -\mathbf{grad}(P + \hat{\mu}\mathbf{H}^0 \cdot \mathbf{h}' + \rho g z) + \hat{\mu}(\mathbf{H}^0 \cdot \mathbf{grad})\mathbf{h}' \quad (8.32)$$

Taking the divergence of this last equation, we obtain

$$\nabla^2(P + \hat{\mu}\mathbf{H}^0 \cdot \mathbf{h}' + \rho g z) = 0$$

Outside the undisturbed region, $\mathbf{h}' = 0$ and $\mathbf{grad}(P + \rho g z) = 0$ according to the condition required for an equilibrium state; consequently,

$$P + \hat{\mu}\mathbf{H}^0 \cdot \mathbf{h}' + \rho g z \quad (8.33)$$

is a solution of the Laplace equation, which reduces to a constant in the undisturbed region. Therefore, this quantity is constant everywhere, and we must consider the momentum equation in its simple form:

or
$$\rho \, \frac{\partial \mathbf{u}'}{\partial t} = \hat{\mu}(\mathbf{H}^0 \cdot \mathbf{grad})\mathbf{h}'$$

$$\rho \, \frac{\partial u_i'}{\partial t} = \hat{\mu} H_j^0 \, \frac{\partial}{\partial x_j} \, h_i'$$

With the rate equation for the magnetic field,

or
$$\frac{\partial \mathbf{h}'}{\partial t} = (\mathbf{H}^0 \cdot \mathbf{grad})\mathbf{u}'$$

$$\frac{\partial h_i'}{\partial t} = H_j^0 \, \frac{\partial}{\partial x_j} \, u_i'$$

which leads to a wave equation. If $\mathbf{H}^0$ is taken to be parallel to the 02 axis, the equations become

$$\frac{\partial \mathbf{h}'}{\partial t} = H^0 \, \frac{\partial \mathbf{u}'}{\partial z} \qquad \frac{\partial \mathbf{u}'}{\partial t} = \hat{\mu} H^0 \, \frac{\partial \mathbf{h}'}{\partial z}$$

$$\frac{\partial^2 \mathbf{h}'}{\partial t^2} = a^2 \, \frac{\partial^2 \mathbf{h}'}{\partial z^2} \qquad \frac{\partial^2 \mathbf{u}'}{\partial t^2} = a^2 \, \frac{\partial^2 \mathbf{u}'}{\partial z^2}$$

with

$$a = H^0 \left(\frac{\hat{\mu}}{\rho} \right)^{1/2} \qquad \text{and} \qquad |\mathbf{H}^0| = H_z^0$$

Corresponding experimental investigations are usually carried out with mercury, the order of magnitude of the magnetic field being 1000 G, in which case, $a = 75$ cm/s. In the center of the earth where the magnetic field is weak, a is less than 1 cm/s. In the solar atmosphere where the density is small, a is very high.

8.7   WAVE SCATTERING DUE TO TURBULENCE

So far, we have been concerned only with deterministic phenomena. We now emphasize a pattern of interaction between a deterministic field such as a sound wave and a turbulent field characterized only by statistical properties.

The wave equation is still assumed to obey a linear equation. The d'alembertian operator is capable of extracting from an equation the part related to wave motion; other terms on the right-hand side of the equation can be interpreted as source terms. The sound wave is assumed to propagate in a medium that exhibits random properties through the pressure of the density terms. In any case, a scalar function $\Psi(\mathbf{x}, t)$ is introduced to account for this phenomenon. We write the equation as

$$\Box\Psi = \nabla^2\Psi - \frac{1}{a^2}\frac{\partial^2\Psi}{\partial t^2} = \text{source term}$$

The source term is derived from the basic equations of motion. For the sake of simplicity, we assume the medium to be homogeneous and the turbulent field to be frozen, which means that the characteristic time of the turbulent motion is small compared with a characteristic time linked to wave propogation. Therefore, time t is introduced only through the Ψ function. We seek a solution by a perturbation procedure.

The source term is introduced in the form

$$\sum P_n(Z)\, D_n(\Psi)$$

where Z is a random variable that characterizes the fluctuations of the medium. The global equation can be written as

$$\Box\Psi = \sum_n P_n(Z)\, D_n(\Psi)$$

D_n is a differential or integral operator applied to the wave function, and $P_n(Z)$ is a function of a group of variables that specify the relevant local properties of the medium. The variables Z, of which temperature is a typical example, are themselves random functions of position and time, which account for the action of turbulent motion in the medium.

The first approximation of the equation is termed Ψ_0, which corresponds to the solution of the following homogeneous equation (the source term has been dropped):

$$\frac{\partial^2\Psi}{\partial t^2} - a^2\frac{\partial^2\Psi}{\partial x^2} = 0 \tag{8.34}$$

Now we consider an incident, monochromatic plane wave, so the first approximation becomes

$$\Psi_0 = A e^{i(\mathbf{k} \cdot \mathbf{x} - \omega t)} \tag{8.35}$$

with

$$|\mathbf{k}| = \frac{\omega}{a_0} = \frac{2\pi}{\lambda}$$

The random motion of the medium is taken into account by means of the function Ψ_1, which represents the first level of this interaction. According to our assumption, the right-hand term of the equation can be written $P(Z) D(\Psi_0)$, so we have to solve the following equation:

$$\frac{\partial^2 \Psi_1}{\partial t^2} = a^2 \frac{\partial^2 \Psi_1}{\partial x_j \, \partial x_j} = \underbrace{A e^{i(\mathbf{k} \cdot \mathbf{x} - \omega t)}}_{\Psi_0} Q(\mathbf{x}) \tag{8.36}$$

The disturbance generated by the random medium at point $\mathbf{y}$ and time $t - \tau$ is traveling at the speed of sound. It can be observed at point $\mathbf{x}$ and time t if we set

$$\tau = \frac{|\mathbf{x} - \mathbf{y}|}{a}$$

where τ is the traveling time, which can be interpreted as a time delay. Finally, the solution is presented in an integral form:

$$\Psi_1 (\mathbf{x}, t) = -\frac{A}{4\pi} \int e^{i(\mathbf{k} \cdot \mathbf{y} - \omega t) + i|\mathbf{k}||\mathbf{x} - \mathbf{y}|} \frac{Q(\mathbf{y})}{|\mathbf{x} - \mathbf{y}|} \, d\mathbf{y}$$

In the exponential term we introduce the time delay τ and the source intensity, taken at time $t = \tau$, so the exponential becomes

$$i \{\mathbf{k} \cdot \mathbf{y} - \omega(t - \tau)\} = i \{\mathbf{k} \cdot \mathbf{y} - \omega t\} + i|\mathbf{k}||\mathbf{x} - \mathbf{y}|$$

A Cauchy kernel $1/|\mathbf{x} - \mathbf{y}|$ appears in the solution.

The disturbance generated by the medium is located at point $\mathbf{y}$ and the observation is made at point $\mathbf{x}$. The sound wave represented by Ψ_1 is the first approximation to the effect of the medium on the incident wave represented by Ψ_0. Ψ_1 has the same form as the wave function appropriate to a volume distribution $\mathcal{V}$ of acoustic sources distributed in the entire space. Each source is of frequency ω and the density at point $\mathbf{y}$ is proportional to $A Q(\mathbf{y})$.

In most cases, $\Psi_0 + \Psi_1$ is a valid approximation of the wave function, provided that

$$\varepsilon \, \frac{(\mathcal{V}^{1/3} L)^{1/2}}{\lambda} \ll 1$$

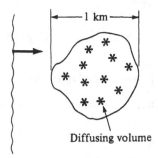

Fig. 8.9   Incident wave on diffusive media

where ε is a measure of the magnitude of the relative variation in the speed of sound, λ is the acoustic wavelength, and L denotes a length scale characteristic of the random property of the medium. Assume that sound waves are scattered by temperature fluctuations characterized by a root mean square value of $1°$ and a length scale $L = 1$ m. If the wavelength of the incident wave is 10 cm, the solution given by the previous equation can be considered valid on the condition that $\mathcal{V}^{1/3} < 1$ km. The size of the diffusing volume (see Fig. 8.9) is thereby limited.

This solution must be presented in a simpler form if the distance from the observation point $\mathbf{x}$ to any point $\mathbf{y}$ belonging to the diffusing volume is great compared with the linear dimension of $\mathcal{V}$. In this case we can write

$$|\mathbf{x} . \mathbf{y}| = \sqrt{\sum_i (x_i - y_i)^2} = \sqrt{\sum_i (x_i^2 - 2x_i y_i + y_i^2)}$$

$$\simeq \sqrt{\sum_i x_i^2 \left(1 - \frac{2x_i y_i}{x_i^2}\right)}$$

$$\simeq |\mathbf{x}| \left[1 - \frac{\mathbf{x} . \mathbf{y}}{|x|^2}\right]$$

$$i|\mathbf{k}||\mathbf{x} - \mathbf{y}| \simeq i|\mathbf{k}||\mathbf{x}| \left[1 - \frac{\mathbf{x} . \mathbf{y}}{|x|^2}\right]$$

for $|\mathbf{y}| \ll |\mathbf{x}|$.

Finally, Ψ_1 can be approximated by

$$\Psi_1(\mathbf{x}, t) = -\frac{A}{4\pi|\mathbf{x}|} \int_{\mathcal{V}} e^{i(\mathbf{k}.\mathbf{y} - \omega t) + i|\mathbf{k}||\mathbf{x}|[1 - (\mathbf{x}.\mathbf{y}/|x|^2)]} Q(\mathbf{y}) \, d\mathbf{y}$$

The following forms can now be derived:

$$\Psi_1(\mathbf{x},\,t) \;=\; -\frac{A}{4\pi|\mathbf{x}|}\, e^{i(|\mathbf{k}||\mathbf{x}|-\omega t)} \int_V e^{i[\mathbf{k}\cdot\mathbf{y}-(|\mathbf{k}||\mathbf{x}|/|\mathbf{x}|^2)\,\mathbf{x}\cdot\mathbf{y}]} Q(\mathbf{y})\;\mathbf{dy}$$

$$=\; -\frac{A}{4\pi|\mathbf{x}|}\, e^{i(|\mathbf{k}||\mathbf{x}|-\omega t)} \int_V e^{i[\mathbf{k}-(|\mathbf{k}|/|\mathbf{x}|)\mathbf{x}]\cdot\mathbf{y}} Q(\mathbf{y})\;\mathbf{dy}$$

We introduce a unit vector $\mathbf{l}$ in connection with the observation point $\mathbf{x}$:

$$\mathbf{l} \;=\; \frac{\mathbf{x}}{|\mathbf{x}|}$$

This unit vector is associated with a wave-number vector $\mathbf{K}$, that is, $\mathbf{K} = \mathbf{k} - |\mathbf{k}|\mathbf{l}$, as shown in Fig. 8.10.

We can now write

$$\Psi_1(\mathbf{x},\,t) \;=\; -\frac{A}{4\pi|\mathbf{x}|}\, e^{i(|\mathbf{k}||\mathbf{x}|-\omega t)} \int_V e^{i\mathbf{K}\cdot\mathbf{y}}\, Q(\mathbf{y})\;\mathbf{dy}$$

$$=\; -\frac{A}{4\pi|\mathbf{x}|}\, e^{i(|\mathbf{k}||\mathbf{x}|-\omega t)}\, \Gamma(\mathbf{K})$$

In these forms the effect of the incident wave is seen from the term

$$e^{i(|\mathbf{k}||\mathbf{x}|-\omega t)}$$

whereas the integral term elucidates the effect of the diffusing medium. Finally, this disturbance appears in the form of a Fourier transform with respect to the wave number $\mathbf{K}$. In other words, the wave number $\mathbf{K}$ plays the role of a parameter.

In the next stage, we need to introduce the quantities responsible for this scattering process in the source term $\Gamma(\mathbf{K})$. The properties of the medium vary randomly as a result of turbulence; therefore, the function $\Gamma(\mathbf{K})$ and consequently $\Psi_1(\mathbf{x},\,t)$ take on values in accordance with each realization. Only the statistical properties of $\Gamma(\mathbf{K})$ and $\Psi_1(\mathbf{x},\,t)$ are relevant. Accordingly, we calculate the mean value of Ψ_1, this being the most significant of these properties. The mean intensity of the scattered wave at position $\mathbf{x}$ expressed as a fraction of the intensity wave is

$$\frac{\Psi_1\,\Psi_1^*}{A^2} \;=\; \frac{\Gamma(K)\,\Gamma^*(K)}{16\pi^2|\mathbf{x}|^2}$$

$$=\; \frac{1}{16\pi^2|\mathbf{x}|^2} \iint Q(\mathbf{y})Q(\mathbf{z})e^{i\mathbf{k}(\mathbf{y}-\mathbf{z})}\;\mathbf{dy}\;\mathbf{dz}$$

The form of this expression for the mean intensity of the scattered wave is worth noting, not only because of its simplicity but also because it

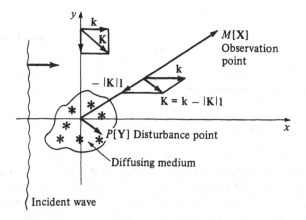

Fig. 8.10 Wave number diagram for diffusive media

reveals the role played by the spectrum of the relevant property of the medium.

Now we set $z = y + \xi$. The distance between two points in the diffusing volume is denoted, and the previous integral becomes

$$\iint_v \overline{Q(y)Q(y + \xi)}\, e^{-i\mathbf{k}\cdot\xi}\, d\mathbf{y}\, d(\mathbf{y} + \xi)$$

If the statistical properties of the medium are almost homogeneous, we can expect the correlation term $\overline{Q(y)Q(y + \xi)}$ to depend on the relative position defined by $\mathbf{y}$ and $\mathbf{y} + \xi$ defined by ξ and not on $\mathbf{y}$, so we can set $\overline{Q(y)Q(y - \xi)} = R(\xi)$. We can easily integrate the previous expression because the variable $\mathbf{y}$ does not appear in the function:

$$\frac{\Psi_1 \Psi_1^*}{A^2} = \frac{1}{16\pi^2|\mathbf{x}|^2} \int_v R(\xi)\, e^{-i\mathbf{k}\cdot\zeta}\, d\xi$$

$\Phi(\mathbf{K})$ is the spectrum of the correlation function $R(\xi)$ with respect to the wave number $\mathbf{K}$.

The last relation emphasizes the combined role of the high wave number $\mathbf{k}$ of the incident wave and that of the observation point by means of a new wave number $\mathbf{K}$.

We can compute the source term for either a random thermal field or a random velocity field. These results are available for the far field but not for the near field, which is a consequence of the assumptions we previously made. This phenomenon is also termed *wave scattering due to turbulence*.

The single-scattering approximation proves to be quite sufficient in many problems of wave propagation in the atmosphere and ocean. This

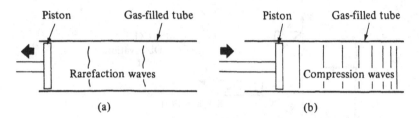

Fig. 8.11   Compression and rarefaction waves

applies to practical problems of the troposphere beyond-the-horizon communication, to scattering from wakes, plumes, aircrafts, or rocket engines, and also to the detection of clear-air turbulence. The detailed theory of single scattering helps in solving the inverse problem; that is, we find the parameters of the medium from the observed characteristics of the scattered field.

8.8 WAVES OF FINITE AMPLITUDE

Although we cannot treat the problem of finite amplitude waves, such as shock waves or denotation waves, in this kind of treatise, we do want to point out some behaviors of shock waves and rarefaction waves because the speed of traveling waves of finite amplitude can be very different from that of a traveling wave of small amplitude.

Compression waves of finite amplitude can coalesce to create shock waves. The two typical behaviors of compression waves (which can coalesce) and rarefaction waves (which cannot coalesce) are worth noting. Such a situation can be explained by making reference to the second principle of thermodynamics.

The entropy of the gas decreases in a rarefaction shock wave, whereas it increases in a compression shock wave. Only this latter effect is possible for an isolated system.

This behavior is easy to understand in another way: In an unsteady plane flow, a disturbance that compresses the medium can lead to a breakdown in the fluid. Imagine that the compression takes place in a gas in a straight tube as the result of the motion of a piston (Fig. 8.11). We will examine only the behavior of the waves traveling in the x direction.

Consider a small rarefaction wave due to a small displacement of the piston in the opposite direction as in Fig. 8.11a. The speed of this wave should not significantly differ from the sound speed in the undisturbed medium at temperature T_0, that is, $a_1 = \sqrt{\gamma R T_0}$. This rarefaction wave introduces a small, negative temperature change ΔT_1, which could be neglected for an isolated wave, but the properties of the medium are altered by a finite number of rarefaction waves. Thus, the second wave

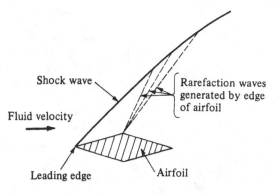

Shock wave

Fluid velocity

Rarefaction waves
generated by edge
of airfoil

Leading edge

Airfoil

Fig. 8.12 Wave interaction past an airfoil

travels in a medium whose temperature is $T_0 - \Delta T_1$, so the sound speed is now reduced to $a_2 = \sqrt{\gamma R(T_0 - \Delta T_1)}$. Because the distance between the two waves increases with time, no coalescence is possible. On the other hand (Fig. 8.11b), a small compression wave introduces a small, positive temperature change, so the speed of the second wave is greater: $a_2 = \sqrt{\gamma R(T_0 + \Delta T_1)}$. Because the distance between these two waves decreases with t, the two waves will coalesce. Thus, we see that a group of compression waves generated at different intervals of time are capable of coalescing. In both cases, the wave systems combine to validate the boundary conditions at the piston wall.

Small waves such as sound waves do not disturb the physical properties of the medium. On the other hand, a finite compression wave locally alters the properties of the medium, and its speed depends on the degree of alteration of these properties.

A group of compression waves that have coalesced can generate a shock wave whose speed is far greater than the speed of sound. The physical properties of the medium are then significantly altered by passing shock waves, a case in point being the temperature, which increases from T_0 to T_1. A crude approximation to the speeds of shock waves is given by the relation.

$$a = \sqrt{\gamma R \left(\frac{T_0 + T_1}{2} \right)} \tag{8.37}$$

with $(T_0 + T_1)/2 > T_0$.

Complex phenomena take place if a group of rarefaction waves interferes with shock waves, as shown in Fig. 8.12. In this particular case, curved shock waves apear, which introduce an entropy gradient in an extended part of the flow. An airfoil can generate this kind of interaction in supersonic flows.

Such problems are dealt with in treatises on high-speed flows.

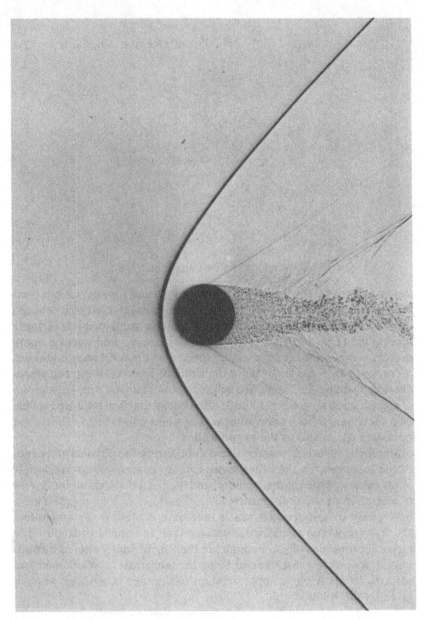

Sphere at $M = 1.53$. A shadowgraph catches a $\frac{1}{2}$-in. sphere in free
flight through air. The flow is subsonic behind the part of the bow
wave that is ahead of the sphere, and over the sphere's surface and
back to 45°. At about 90°, the laminar boundary layer separates from
the sphere through an oblique shock wave and quickly becomes turbulent. The fluctuating wake generates a system of weak disturbances
that merge into the next shock wave. Photograph by A. C. Charters.
Reproduced by permission from M. van Dyke, *An Album of Fluid
Motion*. Parabolic Press, Stanford, California, 1983, p. 164.

9   Fundamentals of Classical Thermodynamics

The science of thermodynamics deals with relations between heat and work and is based on two general laws of nature, the first and second laws of thermodynamics. From these laws it is logically possible to correlate many of the observable properties of matter: specific-heat capacity, coefficient of expansion, vapor pressure, and so on. Thermodynamics is an experimental and empirical science and does not necessarily make assumptions about the nature of matter. Thermodynamic formulas, of course, have the same general validity as the two laws from which they are derived, but the consequence of generality is a restriction in scope.

The most fruitful approach to studying the properties of matter, therefore, combines thermodynamic reasoning with molecular theory. We do not develop such an aspect in this chapter, but the reader may know from other sources how statistical mechanics enables us to go beyond the predictions of pure thermodynamics to gain a deeper insight into the laws of thermodynamics.

Thermodynamic principles provide information in addition to mechanical laws and are therefore used by engineers in the design of subsonic or supersonic aircraft, internal combustion engines, steam engines and turbines, jet engines, and refrigerators, and by chemical engineers in practically every process involving a heat flux or a problem in chemical equilibrium.

9.1.a *Macroscopic and microscopic points of view*

The behavior of a system can be investigated from either a microscopic or a macroscopic point of view. In all cases under consideration and whatever the scale of investigation, the system behaves in a deterministic way. However, the number of parameters can become so large that a statistical approach is the most reasonable. For instance, to treat the problem of a monoatomic gas contained in a 25-mm cube would require at least 6×10^{20} equations. There are two approaches to this problem that can reduce the number of equations and variables to a few: use of statistics (microscopic) and use of classical thermodynamics (macroscopic).

241

The statistical approach deals with "average" values for all particles in a system, based on statistical considerations and probability theory. This is usually done in connection with the modeling of molecular mechanisms, as in kinetic theory and statistical mechanics.

The other approach, the macroscopic point of view of classical thermodynamics is concerned with the average effects of many molecules. These effects can be measured by instruments, though what we really measure is the time-averaged influence of many molecules. For example, consider the pressure a gas exerts on the wall of its container. This pressure can be interpreted from the change of momentum of the molecules as they collide with the wall. The macroscopic point of view is not concerned with the action of individual molecules but only with the time-averaged force on a given area, which can be measured by a pressure gauge. All the discussions in this chapter are presented from a macroscopic point of view, though the microscopic perspective may be included as an aid to understanding the physical processes involved. A macroscopic approach can be accepted if we are concerned with volumes that are very large compared with molecular dimensions and, therefore, with systems that contain many molecules. If we do not try to tackle the motion of individual molecules, we can treat a substance as continuous. From this point of view the *medium* can be called a *continuum*. In most engineering work the assumption of a continuum is valid and convenient, and is in agreement with the microscopic point of view.

From a microscopic point of view an equation of state corresponds to a steady statistical situation concerning the behavior of molecules. In fact, this sort of law can be used in unsteady macroscopic flows. The characteristic time of a molecular field is very small compared with the characteristic time of a flow field, which is, of course, directly connected with the mean velocity gradients. In the usual experimental situations, the smallest turbulent structures are very large compared with the mean free path of the molecules.

The equation of state

$$pv = RT \tag{9.1}$$

is satisfied for an ideal gas and can be applied even though the flow is unsteady. It corresponds to a very simple molecular model. If more complex behaviors are detected for the gas, we must consider a more sophisticated molecular scheme – for example, the van der Waals equation. Proposed in 1873, this equation significantly enlarged the Mariotte law. In any case, the macroscopic and microscopic points of view must go hand in hand.

9.1.b *Properties and state of a substance*

Thermodynamic properties can be divided into two general classes: *intensive properties*, which are independent of the mass, and *extensive properties*, which vary directly with the mass.

Thus, if we divide a quantity of matter in a given state into two equal masses, the intensive properties of each will have the same value as the original quantity of matter; and the extensive properties of each will have one-half the value of the original.

9.1.c *Thermodynamic system*

We present some essential features of compressible fluid flows. Even though we consider only very simple patterns of flow, the interaction between a constitutive law and the second principle of thermodynamics appears to be crucial. Any laws that govern exchanges between two elementary parts of a system have to be consistent with the thermodynamic laws.

If the exchange laws are founded on a first-gradient approximation such as

$$\tau = \mu \frac{du}{dy} \qquad \mathbf{q} = -\lambda \, \mathbf{grad} \, T \qquad (9.2)$$

the problem appears to be coherent; that is, the dissipation function becomes positive, the heat flux is in the direction of decreasing temperature, and matter travels from a domain of high concentration to one of low concentration.

As used in thermodynamics, the term *system* refers to a definite quantity of matter bounded by closed surfaces. The surface may be a real one, such as a tank enclosing a mass of gas, or it may be imaginary, such as the boundary of a certain mass of fluid flowing through a pipe. The boundary surface is not necessarily fixed in either shape or volume. When a fluid expands against a moving piston, the volume enclosed by the boundary surfaces changes. When reference is made to a defined volume at time *t*, it is possible to reintroduce the concept of a "system enclosing a certain mass of gas" by using the lagrangian approach and material derivatives. For a short time, an arbitrary set of particles is followed as they progress. They form a sort of moving system similar to that introduced in the case of a pipe flow.

In fluid mechanics, however, a *system* no longer has a definite quantity of matter. Instead, we must deal with systems in which there is a flow of matter across the boundary. A *control volume* is then defined, as we previously did when we dealt with the continuity or the momentum equation.

Most problems in thermodynamics involve interchange of energy between a given system and other systems that are called the *surroundings* (or *environment*) of the given system. A system exchanges energy with its surroundings by the performance of mechanical work or by heat flow. If conditions are such that no energy interchange can take place, the system is said to be *isolated*. It should be possible to define the thermodynamic state of a system. For instance, a gas in a cylinder is completely specified by its pressure, volume, temperature, and mass. A consequence of this characterization is the introduction of *thermodynamic coordinates*. For the case in which any portion of the cylinder is in the same state as the surrounding domain, we use the concept of an equilibrium state, which we discuss later in detail. The more complex a system is, the more variables must be introduced; in addition to the variables just listed, we may need to specify the values of quantities such as the concentration of a solution, the charge of an electric cell, the polarization in a dielectric, or the area of a surface film.

However, for this chapter, we assume that the quantities whose values determine the state of a system are pressure, density, and temperature. When an isolated system is left to itself and the pressure, temperature, and density are measured at various points throughout the system, we observe that, though these quantities may change with time, the rates of change become smaller and smaller until eventually no further observable change occurs. The final steady state of an isolated system is called a *state of thermodynamic equilibrium*, and we postulate that in such a state the thermodynamic coordinates of a homogeneous system are the same at all points.

Thermodynamics books usually present heterogeneous systems that consist, for example, of a liquid in equilibrium with its vapor. The pressure and the temperature are assumed to be the same at all points, but the density is assumed to be the same only in homogeneous components, be they liquid or gas. Thus, the density is the same at all points occupied by the liquid, and the density of the vapor is the same at all points occupied by the vapor.

When thermodynamic laws are applied to a fluid flow in a one-dimensional frame, we must assume many conditions – after first defining a control volume: The system is assumed to start from a state characterized by the flow entering the control volume and end in a final state characterized by the flow leaving this control volume. The flow is steady. Moreover, molecular properties of the flow are in an equilibrium state at both the entrance and the exit points of the control volume, an assumption that can be justified if characteristic times are compared. This problem is analyzed in the next section.

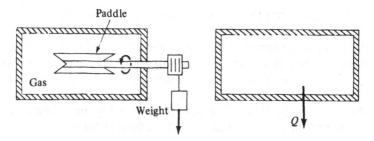

Fig. 9.1 Work done and heat generated by a system

9.1.d *Processes*

Any change in the thermodynamic coordinates of a system is called a *process*. Let us recall the classical definitions of reversible and irreversible processes. A process that is carried out in such a way that at every instant the pressure, temperature, and density of every homogeneous portion of the system remain essentially uniform is called *reversible*. Thus, a reversible process can be defined as a succession of equilibrium states, or at least states that depart only infinitesimally from equilibrium. If there are departures from uniformity, the process is called *irreversible*. Heat transfers with sources at a different temperature from that of the system generate irreversibilities. The significance of the terms reversible and irreversible is evident only in the light of the second law of thermodynamics.

Consider a gas in a cylinder provided with a movable piston. If the piston is pushed down very slowly to minimize internal friction effects and allow time for the pressure and temperature to equalize at all points, the gas is compressed reversibly. The concept of a reversible process must be considered as a reference process. All industrial processes are irreversible because they occur with finite differences of pressure or temperature among parts of a system or between a system and its surroundings.

In thermodynamics, systems often undergo *cycles*. Consider a gas in a container as a system. Let this system go through a cycle that is comprised of two processes. In the first process, work is done on the system by a paddle that turns as a weight is lowered. The second process returns the system to its initial state by transferring heat from the system until the cycle has been completed. Work and heat have been exchanged by the system with the surroundings to restore the system to its initial state. Measurements of work and heat made during one cycle for a variety of systems and for various amounts of work and heat lead to the first law of thermodynamics (Fig. 9.1).

9.2 THE FIRST LAW OF THERMODYNAMICS

We begin by applying the first law of thermodynamics to one-dimensional flows. A control volume is preferentially used if flows are involved. Consider a system that exchanges work and heat with its surroundings. The kinetic energy change is called ΔC, and η is a parameter that characterizes the thermochemical state of the system. The first law of thermodynamics can be stated in the following form:

$$\underset{\text{work}}{\mathcal{T}} + \underset{\text{heat}}{Q} - \Delta C = f(\eta_1, \eta_2) \tag{9.3}$$

where f depends on the initial and final state:

$$f(\eta_1, \eta_2) = \mathcal{E}(\eta_1) - \mathcal{E}(\eta_2) = \Delta \mathcal{E} \tag{9.4}$$

where $\Delta \mathcal{E}$ denotes the internal energy change of the system:

$$\mathcal{T} + Q - \Delta C = \Delta \mathcal{E} \tag{9.5}$$

By convention, energy that enters the system is positive and energy that leaves the system is negative. Two of these four quantities (Q and $\mathcal{E}$) are independent of the system of coordinates, and the other two depend on the system.

When we apply the first law to a fluid particle, we can introduce a new system of coordinates whose origin is at the center of gravity of the particle and whose axes are parallel to those of the galilean frame. Work and kinetic energy are split into two parts:

$$\mathcal{T} = \mathcal{T}^* + \mathcal{T}_M \qquad \Delta C = \Delta C^* + \Delta C_M \tag{9.6}$$

where $\mathcal{T}^*$ and C^*, respectively, refer to work and kinetic energy with respect to the auxiliary frame. The second part of Eq. 9.6 is given by the Koenig theorem.

The first law of thermodynamics can be written as

$$\mathcal{T}^* + \mathcal{T}_M - \Delta C^* - \Delta C_M = \underbrace{\Delta \mathcal{E} - Q}_{\substack{\text{quantities independent} \\ \text{of the frame}}} \tag{9.7}$$

Applying the fundamental law of mechanics to the center of gravity of the fluid particle leads to

$$\mathcal{T}_M + \mathcal{T}_{fM} - \Delta C = 0 \tag{9.8}$$

in which $\mathcal{T}_{fM}$ represents the work done by the friction forces. The centers of gravity of the particles behave as material points. We can also write

$$\mathcal{T}_{fM} < 0 \tag{9.9}$$

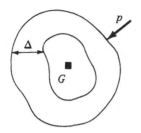

Fig. 9.2   Law of mechanics applied to a fluid particle

If we compare the two preceding equations, we see that

$$\mathcal{T} - \Delta C^* = \Delta \mathcal{E} - q + \mathcal{T}_{fM}$$

Generally, ΔC^* can be neglected, yielding

$$\boxed{\mathcal{T}^* + Q = \Delta \mathcal{E} + \mathcal{T}_{fM}} \qquad (9.10)$$

Considering the motion with respect to the center of gravity G of the fluid particle, an elementary computation gives the amount of work of the pressure forces acting on the fluid surrounding G.

$$\mathcal{T}^* = -p \,\Delta \mathcal{V} \qquad (9.11)$$

or if reference is made to the unit of mass,

$$\mathcal{T}^* = -p \,\Delta v \qquad (9.12)$$

where v is the specific volume of the gas. Generally, we consider a unit mass.

Now we review some fundamental features. We assume the gas to obey the simplest equation of state, $pv = RT$. The effects of the two specific heats C_p and C_v are easy to characterize. In both cases the gas is heated:

   If the volume of the gas is kept constant, no work can be done, which
   leads to

$$Q = \Delta \mathcal{E} = C_v \,\Delta T \qquad (9.13)$$

   If the gas is heated and the pressure is kept constant, we find that

$$Q = \Delta \mathcal{E} + p \,\Delta v \qquad (9.14)$$

The equation of state yields

$$pv = RT \Rightarrow p \,\Delta v + v \,\Delta p = R \,\Delta T \qquad (9.15)$$

When Δp is null, Eq. 9.15 becomes

$$p \,\Delta v = R \,\Delta T \qquad (9.16)$$

and we write

$$p \, \Delta v = R \, \Delta T \tag{9.17}$$

R accounts for the properties of expansion of the gas

$$R = C_p - C_v \tag{9.18}$$

and that yields

$$Q = C_p \, \Delta T \tag{9.19}$$

It is possible to rewrite Eq. 9.19 by using to the variables p and v only:

$$\boxed{Q - \mathcal{T}_{fM} = \frac{C_p}{R} p \, \Delta v + \frac{C_v}{R} v \, \Delta p} \tag{9.20}$$

During the process, we assume that there is no heat exchange with the surroundings ($Q = 0$), and that the effects of friction forces are negligible ($\mathcal{T}_{fM} = 0$). With these two assumptions, the previous equation becomes

$$\frac{C_p}{R} p \, \Delta v + \frac{C_v}{R} v \, \Delta p = 0 \tag{9.21}$$

If we accept the equation of state $pv = RT$ and take into consideration a differential evolution dv and dp, a classical integration can be performed:

$$\frac{p}{p_0} = \left[\frac{\rho}{\rho_0} \right]^{\gamma} \quad \text{with} \quad \rho = \frac{1}{v} \tag{9.22}$$

Using the equation of state to reintroduce the temperature T gives

$$\boxed{\begin{array}{l} \dfrac{p}{p_0} = \left[\dfrac{\rho}{\rho_0} \right]^{\gamma} = \left[\dfrac{T}{T_0} \right]^{\gamma/(\gamma-1)} \\[2ex] \dfrac{T}{T_0} = \left[\dfrac{\rho}{\rho_0} \right]^{\gamma-1} \quad \dfrac{\rho}{\rho_0} = \left[\dfrac{T}{T_0} \right]^{1/(\gamma-1)} \end{array}} \tag{9.23}$$

These relations among the three variables are often used when adiabatic (isolated from the surroundings) and reversible processes come into play.

The use of a simple relation such as $pv^{\gamma} = $ constant is also possible if a balance is obtained between Q and $\mathcal{T}_{fM}$ such that $Q + \mathcal{T}_{fM} = 0$. The amount of heat from friction forces $\mathcal{T}_{fM}$ is precisely balanced by heat exchange with the surroundings. The process is, of course, typically irreversible.

Remark: The effects of friction forces do not appear explicitly in the first law of thermodynamics, which refers to exchanges with the surroundings only. From a thermodynamics point of view, the friction forces and the heat releases that are necessarily associated are both embedded in the system. This internal mechanism corresponds to an alteration of the quality of the energy and not its amount. The role of friction forces will be viewed through the second law of thermodynamics.

9.2.a *Enthalpy*

The elementary work required to move the center of gravity of a fluid particle consists of two elementary parts.

> Because a fluid particle is considered a rigid body, its center of gravity must be moved. This motion is represented by the term $-v\,\Delta p = -\Delta p/\rho$ in the Euler equations.
> A second part is related to the motion of a fluid particle with respect to its center of gravity (or with respect to the auxiliary frame previously introduced). It is represented by the term $-p\,\Delta v$.

Work can be done by the surroundings and introduced into a system. This kind of work is called $\mathcal{T}_u$. This definition will be useful when we use a control volume.

The first law can be written as

$$Q + \mathcal{T}_u - p\,\Delta v - v\,\Delta p = \Delta C + \Delta\mathcal{E} \tag{9.24}$$

It is now convenient to define a new extensive property called the *enthalpy* of the fluid, or its *specific enthalpy* if reference is made to a unit mass. The enthalpy change takes into account both $\Delta\mathcal{E}$ and $\Delta(pv)$. Accordingly, the enthalpy is defined by

$$H = \mathcal{E} + pv$$

Therefore, the first law of thermodynamics can be rewritten as

$$\boxed{Q + \mathcal{T}_u = \Delta C + \Delta H} \tag{9.25}$$

If the equation of state $p = \rho RT$ is introduced,

$$Q + \mathcal{T}_u = \Delta C + \Delta\mathcal{E} + \Delta(pv) \tag{9.26}$$

$$\boxed{Q + \mathcal{T}_u = \Delta C + \Delta H} \tag{9.25}$$

The physical characteristics of a fluid, such as C_p and C_v, are assumed to be constant in the range of our investigations.

9.2.b *Polytropic processes*

Let us recall some features that are useful for technical applications. If heat exchanges and friction losses are given as functions of another quantity such as $v \, \Delta p$, a polytropic coefficient k can be introduced in place of γ.

The fundamental law of mechanics applied to the center of gravity of a fluid particle leads to

$$v \, dp = d\mathfrak{I}_u - d\mathfrak{I}_{fM} \tag{9.27}$$

We can also write

$$\eta \, d\mathfrak{I}_u = v \, dp \tag{9.28}$$

and

$$d\mathfrak{I}_{fM} = d\mathfrak{I}_u(1 - \eta) \tag{9.29}$$

The heat exchanges are given with reference to $d\mathfrak{I}_u$:

$$d \, Q = \mathscr{E} \, d\mathfrak{I}_u \tag{9.30}$$

We can write

$$\mathscr{E} \, d\mathfrak{I}_u + (1 - \eta) \, d\mathfrak{I}_u = \frac{C_p}{R} p \, dv + \frac{C_v}{R} v \, dp \tag{9.31}$$

$$\frac{v \, dp}{\eta}(1 - \eta + \mathscr{E}) = \frac{C_p}{R} p \, dv + \frac{C_v}{R} v \, dp \tag{9.32}$$

$$\frac{C_p}{R} p \, dv + v dp \left[\frac{C_v}{R} - \frac{1 - \eta + \mathscr{E}}{\eta} \right] = 0 \tag{9.33}$$

Integrating yields

$$k = \frac{\gamma}{\gamma \left[1 - \dfrac{1 + \mathscr{E}}{\eta} \right] + \dfrac{1 + \mathscr{E}}{\eta}} \tag{9.34}$$

where k is the polytropic coefficient we must introduce in the following expression:

$$pv^k = \text{constant} \tag{9.35}$$

$$k = \gamma \quad \text{if} \quad \mathscr{E} = 0 \quad \text{and} \quad \eta = 1; \mathscr{E} + 1 = \eta$$

$$\text{or if} \quad 1 + \mathscr{E} = \eta \Rightarrow \mathscr{E} = \eta - 1 \leq 0$$

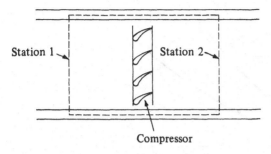

Fig. 9.3 Compressor or turbine

In this last case the amount of heat that must be extracted from the system is equal to the heat produced by the friction forces. Generally

$$k \neq \gamma \qquad (9.36)$$

9.2.c *Industrial application*

The application we examine is a compressor (or turbine) (Fig. 9.3). Reference is made to a control volume. The system is assumed to be thermally isolated from the surroundings. Moreover, the change of kinetic energy between stations 1 and 2 is assumed to be negligible. Equation 9.25 leads to

$$\mathcal{T}_u = \Delta H = C_p T_2 - C_p T_1 = C_p T_1 \left[\frac{T_2}{T_1} - 1 \right] \qquad (9.37)$$

Through the compressor, the fluid goes from pressure level p_1 to pressure level p_2. If the internal losses are negligible, the temperature ratio T_2/T_1 is simply connected to the pressure ratio p_2/p_1, which leads to

$$\frac{T_2}{T_1} = \left[\frac{p_2}{p_1} \right]^{(\gamma - 1)/\gamma} \qquad (9.38)$$

and

$$\mathcal{T}_u = C_p T_1 \left\{ \left[\frac{p_2}{p_1} \right]^{(\gamma - 1)/\gamma} - 1 \right\} \qquad (9.39)$$

in which $\mathcal{T}_u$ is given for a unit mass of fluid crossing the compressor. If internal losses are taken into account, the same pressure ratio is obtained for a change of enthalpy $H_1 - H_2'$ with $H_2' > H_2$:

$$\mathcal{T}_u' = C_p T_1 \left[\frac{T_2'}{T_1} - 1 \right] \qquad (9.40)$$

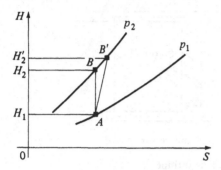

Fig. 9.4 Mollier diagram for an ideal (path AB) and real (path AB') compressor

We define the efficiency of the compressor by making reference to an ideal compressor:

$$\eta = \frac{\mathcal{T}_u}{\mathcal{T}_u'} = \frac{(T_2/T_4) - 1}{(T_2'/T_1) - 1} = \eta \tag{9.41}$$

The real and ideal processes are represented on the Mollier diagram (Fig. 9.4). The ideal path is AB and the real path is AB'. In both cases, the pressure change is fixed $(p_2 - p_1)$.

If the change of pressure is small, the previous expression for $\mathcal{T}_u$ can be rearranged, and for a mass flow rate q_m we can write

$$\mathcal{T}_u = q_m \, C_p \, T_1 \left\{ \left[\frac{p_1 + \Delta p}{p_1} \right]^{(\gamma - 1)/\gamma} - 1 \right\}$$

$$= q_m \, C_p \, T_1 \left\{ \left[1 + \frac{\Delta p}{p_1} \right]^{(\gamma - 1)/\gamma} - 1 \right\} \tag{9.42}$$

where $\Delta p/p_1$ is assumed to be small with respect to unity. Using a power series for

$$\left[1 + \frac{\Delta p}{p_1} \right]^{(\gamma - 1)/\gamma}$$

leads to

$$\mathcal{T}_u \simeq \rho_1 \, q_{v_1} \, C_p \, T_1 \frac{\gamma - 1}{\gamma} \frac{\Delta p}{p_1}$$

$$\simeq \frac{\rho_1}{p_1} q_{v_1} \, C_p \, T_1 \frac{\gamma - 1}{\gamma} \Delta p = \frac{1}{RT_1} q_{v_1} \, C_p \, T_1 \frac{\gamma - 1}{\gamma} \Delta p$$

$$\simeq q_v \, \Delta p \tag{9.43}$$

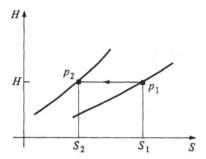

Fig. 9.5 Mollier diagram for an isothermal compression

9.2.d *Isothermal process*

For an isothermal compression, the amount of mechanical work introduced inside the control volume is equal to the amount of heat extracted,
so we have

$$\underbrace{\mathcal{T}_u}_{0} + Q = \Delta H + \delta C \tag{9.44}$$

The corresponding path drawn on the Mollier diagram is shown in Fig.
9.5. The amount of energy remains constant while its quality improves,
which is translated into a decrease of entropy.

9.3 THE SECOND LAW OF THERMODYNAMICS

The first principle refers to the amount of energy, whereas the second
principle considers the quality of this energy.

9.3.a *The Clausius inequality*

The Clausius inequality is a relation between the temperature of an
arbitrary number of heat reservoirs and the quantities of heat extracted
from or absorbed by them. This extraction on absorption occurs when
some working substance is carried through an arbitrary cyclic process
in the course of which it interchanges heat with the reservoirs.

Clausius postulated that if ΔQ_j is the amount of heat conveyed from
the source to the system and if T_j is the temperature of this source, the
quantity

$$\sum_j \frac{\Delta Q_j}{T_j} \leq 0 \tag{9.45}$$

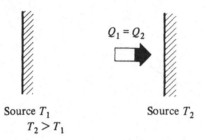

Source T_1 Source T_2
$T_2 > T_1$

Fig. 9.6   Flow of heat from a reservoir

The amount of heat is an algebraic value. By convention it is positive if heat is transferred from the sources to the system and negative in the opposite case. If the system is in contact with continuously distributed sources, the summation becomes an integral:

$$\int \frac{dQ}{T} \leq 0 \tag{9.46}$$

Both Eq. 9.45 and Eq. 9.46 express the Clausius inequality. The quantities Q and dQ are considered positive when heat is given up by a reservoir to a system and negative in the opposite case. Note that the two inequalities make no statements about the temperature of the system under consideration. The only temperatures considered are those of the reservoirs. To illustrate the meaning of these inequalities we give three examples.

First, consider the flow of heat by conduction from a reservoir at a temperature T_2 to a reservoir at a lower temperature T_1 (Fig. 9.6). In this very simple case no system is involved, other than the material that conducts heat from one source to the other. The two quantities of heat Q_2 and Q_1 are necessarily equal in amount; but Q_2 is positive whereas Q_1 is negative.

Let $T_2 = 400$ K, $T_1 = 200$ K, $Q_2 = 800$ J, and $Q_1 = -800$ J. Under these conditions, we see that $\Sigma\, Q/T$ is less than zero because

$$\Sigma \frac{Q}{T} = \frac{800}{400} + \frac{(-800)}{200} = -2 \text{ J/K}$$

The second example is an engine operating between two reservoirs at 400 K and 200 K. The engine is assumed to take 800 J from the high-temperature reservoir and reject 600 J to the low-temperature reservoir (Fig. 9.7). Thus, the system converts 200 J into work. For this system,

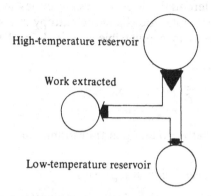

High-temperature reservoir

Work extracted

Low-temperature reservoir

Fig. 9.7 Engine operating between two reservoirs

Eq. 9.45 leads to

$$\Sigma \frac{Q}{T} = \frac{800}{400} + \frac{(-600)}{200} = -1 \text{ J/K}$$

and again $\Sigma \, Q/T$ is negative.

As a third example we present the *limiting cycle* of Carnot. This Carnot engine is assumed to take 800 J from the high-temperature reservoir (800°) and to reject 400 J to the low-temperature reservoir (400°). Thus, this system converts 400 J into work; so we have

$$\Sigma \frac{dQ}{T} = \frac{800}{400} - \frac{(-400)}{200} = 0$$

This limiting cycle is an optimum. It is not possible to do better without transgressing the second law of thermodynamics. In the Carnot cycle, the heat exchanges with the sources are proportional to the temperatures of the sources:

$$\frac{Q_1}{Q_2} = \frac{T_1}{T_2} \qquad (9.47)$$

The amount of heat converted into work is equal to $Q_2 - Q_1$.

9.3.b *Entropy*

The Clausius principle states that, for a reversible cycle, $\int dQ/T$ is null; that is, dQ/T is an exact total differential. The integral $\int dQ/T$ is independent of the path that represents the intermediate states going from state A to state B. In other words, dQ is not an exact total differential; but for a reversible cycle $1/T$ is an integrating factor with which to

generate an exact total differential where reversible cycles are concerned. As any other state variable, the change of entropy of a system
going from state A to state B in a reversible way is defined by the
integral

$$\int_A^B \frac{dQ}{T} = S_B - S_A \qquad (9.48)$$

This change of entropy is relative to an open transformation.

9.3.c *Causes of irreversibility of a process*

Irreversibilities can be introduced either during heat transfer from the
source to the system or during the process itself. Friction phenomena
are usually responsible for this sort of irreversibility. Irreversibilities due
to heat transfer can be taken into account or neglected. They can be
ignored if reference is made to the temperature of the system and not
to that of the source. If isolated processes are considered, irreversibilities
can be introduced by intrinsic phenomena such as friction losses.

9.3.d *Extension of the concept of entropy to irreversible processes*

Consider a natural (and therefore irreversible) process leading from a
representative point A to another point B. In Fig. 9.8 this evolution
(call it process I) is represented by a dotted line. At points A and B we
are able to define the states of the system if we assume that the system
goes from an equilibrium state to another. Then, to this irreversible
process let us associate a reversible one that leads from point B to point
A (call it process II). It may involve interchanges of heat and work with
the surroundings. It is represented in the figure by a solid line. Taken
together, processes I and II constitute a cycle, which as a whole is
irreversible because process I is irreversible. Hence, from the Clausius
inequality, we can write

$$\oint \frac{dQ}{T} < 0 \qquad (9.49)$$

Splitting the integral into two parts yields

$$\underbrace{\int_A^B \frac{dQ}{T}}_{\text{irreversible}} + \underbrace{\int_B^A \frac{dQ}{T}}_{\text{reversible}} < 0 \qquad (9.50)$$

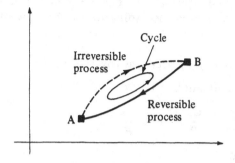

Fig. 9.8   Reversible process

Entropy can be introduced for the reversible part of the system; thus,

$$\int_A^B \frac{dQ}{T} + S_A - S_B < 0 \qquad (9.51)$$

If the irreversible part of the process corresponds to an isolated process, $dQ = 0$; then

$$\boxed{S_A < S_B} \qquad (9.52)$$

The entropy of an isolated system increases in every natural (i.e., irreversible) process.

9.3.e   *Comments on the second law of thermodynamics*

The entropy change in a system determines the direction in which a given process, consistent with the first law, can go. The physical chemist is chiefly concerned with this aspect of entropy: Will two substances react chemically or not? If they do, the reaction should be associated with an increase of entropy. However, even though the entropy might decrease if the reaction were to take place at one temperature and pressure, it still could increase at other values of temperature and pressure. Hence, a knowledge of the entropies of substances as functions of temperature and pressure is important in determining whether chemical reactions will occur.

The mechanical engineer is interested in entropy changes for different reasons. From her point of view, something has been lost when an irreversible process comes into play in a compressor or a turbine. What is lost, of course, is not energy but opportunity, the opportunity to convert internal energy to mechanical energy.

For an isolated system entropy can never be destroyed. Energy can neither be created nor destroyed, according to the first law of thermodynamics.

9.3.f   *The entropy of a perfect gas* (pv = RT)

We use the fundamental equation

$$dQ = \frac{C_p}{R} p \, dv + \frac{C_v}{R} v \, dp \qquad (9.53)$$

Multiplying by $1/T$ we obtain

$$\frac{1}{T} dQ = \frac{C_p}{R} \frac{p \, dv}{T} + \frac{C_v}{R} \frac{v \, dp}{T} \qquad (9.54)$$

When reference is only to the gas, T is the temperature of the gas (or of the sources if the process is reversible, which is assumed*). This leads to

$$T = \frac{pv}{R} \qquad (9.55)$$

$$S_B - S_A = C_p L \left| \frac{v_B}{v_A} \right| + C_v L \left| \frac{p_B}{p_A} \right|$$

$$= C_v L \left\{ \left(\frac{v_B}{v_A} \right)^{C_p/C_v} \times \frac{p_B}{p_A} \right\} \qquad (9.56)$$

$$\boxed{ e^{(S_B - S_A)/C_v} = \frac{p_B \, v_B^{\gamma}}{p_A \, v_A^{\gamma}} } \qquad (9.57)$$

Remark: The role of irreversibilities can be illucidated by means of a typical case, starting from the equation

$$dQ - d\mathcal{T}_{fM} = \frac{C_p}{R} p \, dv + \frac{C_v}{R} v \, dp \qquad (9.58)$$

Multiplying this equation by $1/T$, we note that the term on the right-hand side becomes an exact total differential. T is taken to be the temperature of the fluid, not that of the reservoirs. Equation 9.58 then becomes

$$\frac{dQ}{T} - \frac{d\mathcal{T}_{fM}}{T} = \frac{C_p}{R} \frac{p \, dv}{T} - \frac{C_v}{R} \frac{v \, dp}{T} = \frac{d\mathcal{T} - p \, dv}{T} \qquad (9.59)$$

---

* In this case the temperature of the reservoirs and the temperature of the gas may be only slightly different.

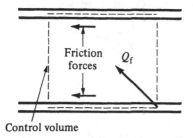

Control volume

Fig. 9.9   Isolated system

This can be rewritten as

$$\frac{dQ}{T} = \frac{d\mathcal{I} + p \, dv}{T} + \frac{d\mathcal{I}_{fM}}{T} \tag{9.60}$$

Even though we consider only the intrinsic irreversibilities of the process, dQ/T is not an exact total differential; thus, we have

$$\frac{dQ}{T} = dS + \frac{d\mathcal{I}_{fM}}{T} \tag{9.61}$$

$$\underbrace{\left[\frac{dQ}{T}\right]}_{\substack{\text{irreversible} \\ \text{process}}} = \underbrace{\left[\frac{dQ}{T}\right]}_{\substack{\text{reversible} \\ \text{process}}} + \frac{d\mathcal{I}_{fM}}{T} \tag{9.62}$$

With our convention, $d\mathcal{I}_{fM}$ is negative, so it appears that the reversible process requires a larger amount of heat from the reservoirs than the irreversible process. Friction losses generate an amount of heat inside the medium.

> We emphasize the two causes of irreversibilities:
> Heat-transfer irreversibilities play a role if the temperature of the reservoirs is different from that of the fluid. The increase of entropy due to this cause is denoted as dS_e.
> Intrinsic irreversibilities are a consequence of friction forces. The increase of entropy due to this cause is termed dS_i.

Thus, we have

$$dS = dS_e + dS_i \tag{9.63}$$

which, for an isolated process (Fig. 9.9), becomes

$$dS_e = 0 \quad \text{and} \quad dS = dS_i$$

The entropy of this natural isolated system increases, which leads to

$$dS_i > 0 \Rightarrow -\frac{dT_{fM}}{T} > 0 \qquad (9.64)$$

and

$$\mathcal{T}_{fM} < 0 \qquad (9.65)$$

9.3.g *Applications: Case of a perfect gas* (pv = RT)

Influence of friction forces and heat exchanges on the properties of a flow: The first law of thermodynamics can be written as

$$d\mathcal{T}_u + dQ = \frac{dV^2}{2} + dH = aH_0 = C_p \, dT_0 \qquad (9.66)$$

In so doing we introduce a new quantity H_0 that accounts for both the enthalpy of a fluid and its kinetic energy. It is a useful quantity for applications. On the other hand, H can be introduced in the expression of entropy to yield

$$dS = \frac{d\mathcal{T} - p \, dv}{T} = \frac{d\mathcal{T} + d(pv) - v \, dp}{T} = \frac{dH - v \, dp}{T} \qquad (9.67)$$

Because we have a reversible process, we can write

$$S = S_0 \quad \text{and} \quad dS = dS_0 \qquad (9.68)$$

The enthalpy H_0 is that of the fluid brought from speed V to rest by an isolated process; and p_0 is the pressure of fluid at rest, brought to rest through an isolated and reversible process. Accordingly, Eq. 9.68 leads to

$$dS_0 = \frac{dH_0 - v_0 \, dp_0}{T_0} = dS_e + dS_i = dS \qquad (9.69)$$

where $dS_e = \dfrac{dQ}{T}$, and T is assumed to be the temperature of the fluid, which means that heat exchanges are taken into account but not the irreversibilities that are linked to heat transfer from the reservoirs to the fluid. Finally, we obtain

$$dH_0 = v_0 \, dp_0 + T_0 \, dS_e + T_0 \, dS_i \qquad (9.70)$$

because we set

$$dS_e = \frac{dQ}{T} \qquad (9.71)$$

yielding

$$d\mathcal{T}_u + T\,dS_e = dH_0 = v_0\,dp_0 + T_0\,dS_e + T_0\,dS_i \qquad (9.72)$$

and

$$v_0\,dp_0 - d\mathcal{T}_u + [T_0 - T]\,dS_e + T_0\,dS_i = 0 \qquad (9.73)$$

Examining the first equation,

$$d\mathcal{T}_u + T\,dS = dH_0 = C_p\,dT_0 \qquad (9.74)$$

we can state that T_0 is dependent only on the exchanges (heat and work) with the surroundings. As for the second equation,

$$v_0\,dp_0 - d\mathcal{T}_u + [T_0 - T]\,dS_e + T_0\,dS_i = 0 \qquad (9.75)$$

which refers to the pressure of fluid p_0. If

$$d\mathcal{T}_u = 0 \qquad \text{and} \qquad dS_e = 0 \qquad (9.76)$$

we have

$$v_0\,dp_0 = -T_0\,dS_i < 0 \qquad (9.77)$$

and accordingly, the pressure p_0 decreases (loss of total head).

If the surroundings leave work in the system, p_0 increases. If the surroundings leave heat in the system,

$$C_p\,dT_0 = dQ \qquad (9.78)$$

and H_0 increases where p_0 decreases:

$$v_0\,dp_0 = -[T_0 - T]\,dS_e \qquad (9.79)$$

$$T_0 - T > 0 \Rightarrow -[T_0 - T]\,dS_e < 0 \qquad (dQ > 0) \qquad (9.80)$$

If the speed of the fluid is small, $(T_0 - T)$ is also small, and p_0 is not significantly affected.

Shock waves in high-speed flow; normal shock waves: We use a control volume as represented in Fig. 9.10. The unknown functions are p_2, ρ_2, T_2, and V_2. The system is assumed to be isolated. The following four equations can be written:

The momentum equation

$$p_1 + \rho V_1^2 = p_2 + \rho_2 V_2^2 \qquad (9.81)$$

The continuity equation

$$\rho_1 V_1 = \rho_2 V_2 \qquad (9.82)$$

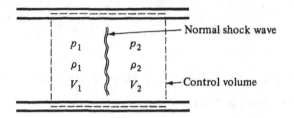

Fig. 9.10   Normal shock wave

The first law of thermodynamics

$$H_1 + \frac{V_1^2}{2} = H_2 + \frac{V_2^2}{2} \qquad (9.83)$$

The equation of state

$$p_1 = \rho_1 RT_1 \qquad p_2 = \rho_2 RT_2 \qquad (9.84)$$

We now introduce a new parameter, the Mach number M. For high-speed flows, we usually refer to the speed of sound a. We show that

$$a_1 = \sqrt{\gamma RT_1} \qquad a_2 = \sqrt{\gamma RT_2} \qquad (9.85)$$

and

$$M_1 = \frac{V_1}{a_1} \qquad M_2 = \frac{V_2}{a_2} \qquad (9.86)$$

After some algebraic computations we find two solutions, the first of which is trivial:

$$p_2 = p_1 \qquad \rho_2 = \rho_1 \qquad T_2 = T_1 \qquad V_2 = V_1 \qquad (9.87)$$

The second solution corresponds to a discontinuity in the flow (at least in the framework of these assumptions which neglect, for instance, the thermal conductivity of the gas); thus, we have

$$\frac{p_2}{p_1} = \frac{2\gamma M_1^2}{\gamma + 1} + \frac{\gamma - 1}{\gamma + 1} \qquad (9.88)$$

$$\frac{V_2}{V_1} = \frac{2}{\gamma + 1}\frac{1}{M_1^2} + \frac{\gamma - 1}{\gamma + 1} \qquad (9.89)$$

$$\frac{\rho_2}{\rho_1} = \frac{(\gamma - 1) M_1^2}{(\gamma - 1) M_1^2 + 2} \qquad (9.90)$$

$$\frac{T_2}{T_1} = \frac{2\gamma M_1^2 (\gamma - 1)}{(\gamma - 1)^2} - \frac{6\gamma - \gamma^2 - 1}{(\gamma + 1)^2}$$

$$- \frac{2\gamma - 2}{(\gamma + 1)^2 M_1^2} \tag{9.91}$$

$$S_1 - S_2 = C_p L \left| \frac{2}{\gamma + 1} \frac{1}{M_1^2} + \frac{\gamma - 1}{\gamma + 1} \right|$$

$$+ C_v L \left| \frac{2\gamma M_1^2}{\gamma + 1} - \frac{\gamma - 1}{\gamma + 1} \right| \tag{9.92}$$

The second law of thermodynamics does not yet intervene. However, two cases must be distinguished.

> If $M_1 < 1$, there are expansion shock waves that are predicted with $S_2 - S_1 < 0$, which is impossible.
> If $M_1 > 1$ there are compression shock waves that are also determined with $S_2 - S_1 > 0$, which is possible.

It is easy to show that elementary waves of compression tend to coalesce to produce a shock wave whenever unsteady motions occur. Oblique shock waves can be considered as the result of a normal shock in the presence of a component of the flow that is parallel to the shock front. This component of the flow is unaltered, giving (see figure):

$$\rho_1 V_{n,1} = \rho_2 V_{n,2} \tag{9.93}$$

$$V_{T_1} = V_{T_2} \tag{9.94}$$

$$\tan \alpha_1 = \frac{V_{T_1}}{V_{n_1}} \qquad \tan \alpha_2 = \frac{V_{T_2}}{V_{n_2}} \tag{9.95}$$

$$\frac{\tan \alpha_1}{\tan \alpha_2} = \frac{V_{n_2}}{V_{n_1}} = \frac{\rho_1}{\rho_2} \tag{9.96}$$

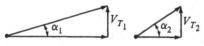

In supersonic flows the locations of the shock are as represented in Fig. 9.11.

Several patterns of interaction can be investigated, such as

> intersection of shocks, which can generate vortex lines
> reflexion of shock waves on a wall
> intersection of shock waves with expansion waves

9.4   THE ENERGY EQUATION AT A POINT

The system consists of a set of fluid particles that are followed during their motion. This approach requires us to enlarge the concept of material derivatives.

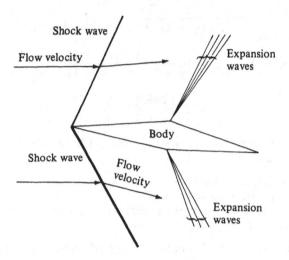

Fig. 9.11   Location of shock waves in supersonic flow

9.4.a *Material derivatives; some extensions*

We treat the problem of material derivatives in an intuitive way. Let us consider a function involved in the integral

$$F = \int_{\mathcal{V}(t)} f(\mathbf{x}, t) \, d\mathcal{V} \tag{9.97}$$

The volume $\mathcal{V}(t)$ is a function of t. The evolution of the elementary volume (Fig. 9.12) depends on the divergence operator; if $f(\mathbf{x}, t) \, d\mathcal{V}$ is considered a simple product, we can assume that the material derivative of the product $f \, d\mathcal{V}$ is given by

$$\frac{df}{dt} \, d\mathcal{V} + f \operatorname{div} \mathbf{V} \, d\mathcal{V} \tag{9.98}$$

which becomes

$$\frac{dF}{dt} = \int_{\mathcal{V}} \left\{ \frac{df}{dt} + f \operatorname{div} \mathbf{V} \right\} d\mathcal{V} \tag{9.99}$$

The equivalent forms are

$$\frac{dF}{dt} = \int_{\mathcal{V}} \left\{ \frac{\partial f}{\partial t} + \underbrace{\mathbf{V} \cdot \operatorname{\mathbf{grad}} f + f \operatorname{div} \mathbf{V}}_{\operatorname{div}(f\mathbf{V})} \right\} d\mathcal{V} \tag{9.100}$$

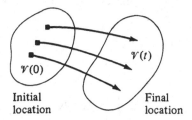

Initial
location

Final
location

Fig. 9.12   Evolution of an elementary volume

$$= \int_{\mathcal{V}} \left\{ \frac{\partial f}{\partial t} + \text{div}(f\mathbf{V}) \right\} d\mathcal{V} \tag{9.101}$$

$$= \int_{\mathcal{V}} \frac{\partial f}{\partial t} d\mathcal{V} + \int_{S} [f\mathbf{V}] \cdot \mathbf{n} \, dS \tag{9.102}$$

If we are concerned with a product such as ρf, we obtain

$$\frac{d}{dt} \int_{\mathcal{V}} \rho f \, d\mathcal{V} = \int_{\mathcal{V}} \rho \frac{df}{dt} d\mathbf{V} + \int_{\mathcal{V}} f \underbrace{\left\{ \frac{df}{dt} + \rho \, \text{div} \, \mathbf{V} \right\}}_{= \, 0} d\mathcal{V} \tag{9.103}$$

$$= \int_{\mathcal{V}} \rho \frac{df}{dt} d\mathcal{V} \tag{9.104}$$

9.4.b   *The energy equation*

We start again from the equation

$$d\mathcal{T} + dQ = d\mathcal{E} + dC = de_0 \tag{9.105}$$

with

$$e_0[\mathbf{x}, t] = \mathcal{E}[\mathbf{x}, t] + \frac{V^2}{2} \tag{9.106}$$

For the set of particles belonging to the volume $\mathcal{V}$, we obtain

$$\mathcal{T}_0[\mathbf{x}, t] = \int_{\mathcal{V}(t)} \rho(\mathbf{x}, t) \, e_0(\mathbf{x}, t) \, d\mathcal{V} \tag{9.107}$$

The actions of the surrounding fluid on the set of particles included in the volume $\mathcal{V}$ are the unique forces whose work we will take into account, which leads to

$$d\mathcal{T} = dt \int_{S(t)} \mathbf{T} \cdot \mathbf{V} \, dS \tag{9.108}$$

The heat flux crossing the surface S is equal over the time interval dt to

$$dQ = dt \int_{S(t)} -\mathbf{q} \cdot \mathbf{n} \, dS \qquad (9.109)$$

Finally, we find that

$$\frac{d\mathcal{E}_0}{dt} = \int_{S(t)} -q_i n_i \, dS + \int_{S(t)} T_i V_i \, dS \qquad (9.110)$$

with

$$T_i = T_{ij} n_j \qquad (9.111)$$

Reintroducing the integral form of $\mathcal{E}_0$ gives

$$\int_{V(t)} \rho \frac{de_0}{dt} \, dV = \int_{V(t)} -\frac{\partial q_i}{\partial x_i} \, dV + \int_{V(t)} \frac{\partial (T_{ij} V_i)}{\partial x_j} \, dV \qquad (9.112)$$

At a given point we have

$$\rho \frac{de_0}{dt} = -\frac{\partial q_i}{\partial x_i} + \frac{\partial}{\partial x_j}(T_{ij} V_i) \qquad (9.113)$$

$$\boxed{\rho \frac{d\mathcal{E}}{dt} + \rho \frac{d}{dt} \frac{V^2}{2} = -\frac{\partial q_i}{\partial x_i} + \frac{\partial}{\partial x_j}(T_{ij} V_i)} \qquad (9.114)$$

Comparisons can be made with the consequences of the momentum equation, taken in the form

$$\rho \frac{dV_i}{dt} = \frac{\partial T_{ij}}{\partial x_j} \qquad (9.115)$$

Multiplying by V_i yields

$$\rho \frac{d}{dt} \frac{V^2}{2} = V_i \frac{\partial T_{ij}}{\partial x_j} \qquad (9.116)$$

By subtraction we find

$$\boxed{\rho \frac{d\mathcal{E}}{dt} = -\frac{\partial q_i}{\partial x_i} + T_{ij} \frac{\partial V_i}{\partial x_j}} \qquad (9.117)$$

This relation is extracted from the global equation and from the momentum equation applied to the center of gravity of a fluid particle. Accordingly, this relation must be considered to be the energy equation with respect to the center of gravity of a fluid particle.

9.4.c *Introduction of a constitutive law*

The previous relations do not anticipate the behavior of the fluid material. At this stage we must formulate a constitutive law. Let us assume the fluid to be a Newtonian fluid. Hence, a relation is assumed to exist between T_{ij} and D_{ij}:

$$T_{ij} = \{-p + \lambda \text{ div } \mathbf{V}\} \, \delta_{ij} + 2\mu \, D_{ij} \tag{9.118}$$

with

$$D_{ij} = \frac{1}{2} \left\{ \frac{\partial V_i}{\partial x_j} + \frac{\partial V_j}{\partial x_i} \right\} \tag{9.119}$$

Introducing T_{ij} into the equation, we have

$$\rho \frac{d\mathscr{E}}{dt} = -\frac{\partial q_i}{\partial x_i} + T_{ij} \frac{\partial V_i}{\partial x_j} \tag{9.120}$$

$$\boxed{\rho \frac{d\mathscr{E}}{dt} = -\frac{\partial q_i}{\partial x_i} - p \text{ div } \mathbf{V} + \lambda [\text{div } \mathbf{V}]^2 + 2\mu \, D_{ij} D_{ij}} \tag{9.121}$$

Equation 9.121 accounts for the symmetry of the tensor D_{ij}, which leads to

$$\sum_{ij} T_{ij} \frac{\partial V_i}{\partial x_j} = T_{ij} D_{ij} \tag{9.122}$$

The dissipation function that appears in the energy equation with respect to the center of gravity appears as an auxiliary heat source.

For a Newtonian fluid this function becomes

$$\Phi = \lambda (\text{div } \mathbf{V})^2 + 2\mu \, D_{ij} D_{ij} \tag{9.123}$$

In this form, λ is not exclusively connected with the expansion of the fluid; a part of this expansion is embedded in the last term. Therefore, we have to make some rearrangements to emphasize this expansion process. Let us introduce a deviatoric tensor d_{ij}.

$$d_{ij} = D_{ij} - \frac{1}{3} D_{kk} \, \delta_{ij} = D_{ij} - \frac{1}{3} (\text{div } \mathbf{V}) \, \delta_{ij} \tag{9.124}$$

$$D_{ij} = d_{ij} + \frac{1}{3} (\text{div } \mathbf{V}) \, \delta_{ij} \tag{9.125}$$

$$\underbrace{\phantom{D_{ij} = d_{ij}}}_{\text{global straining}} \qquad \underbrace{\phantom{+ \frac{1}{3} (\text{div } \mathbf{V}) \, \delta_{ij}}}_{\text{expansion process}}$$

The dissipation function can be rewritten as

$$\Phi = \lambda (\text{div } \mathbf{V})^2 + 2\mu \left\{ d_{ij} + \frac{1}{3} \text{div } \mathbf{V} \delta_{ij} \right\} \left\{ d_{ij} + \frac{1}{3} \text{div } \mathbf{V}_{\cdot ij} \right\} \tag{9.126}$$

and taking into account the relations

$$d_{ii} = 0 \Rightarrow \frac{1}{3} \operatorname{div} \mathbf{V} \, \delta_{ij} \, d_{ij} = \frac{1}{3} \operatorname{div} \mathbf{V} \, d_{ii} = 0 \tag{9.127}$$

and

$$\left\{ \frac{1}{3} (\operatorname{div} \mathbf{V}) \, \delta_{ij} \right\} \frac{1}{3} (\operatorname{div} \mathbf{V}) \, \delta_{ij} = \frac{1}{9} (\operatorname{div} \mathbf{V})^2 \, \delta_{ij} = \frac{1}{3} (\operatorname{div} \mathbf{V})^2 \tag{9.128}$$

we finally have

$$\Phi = \lambda (\operatorname{div} \mathbf{V}) + 2\mu \, d_{ij} \, d_{ij} + \frac{2\mu}{3} (\operatorname{div} \mathbf{V}) \tag{9.129}$$

$$\boxed{\Phi = \left(\lambda + \frac{2\mu}{3} \right) (\operatorname{div} \mathbf{V})^2 + 2\mu \, d_{ij} \, d_{ij}} \tag{9.130}$$

If we state that the dissipation function has to be positive whatever the process under consideration, we must examine the following two cases.

The density of the fluid is assumed to be constant. Thus,

$$\Phi = 2\mu \, d_{ij} \, d_{ij} > 0 \tag{9.131}$$

and

$$\mu > 0 \tag{9.132}$$

The fluid is subjected only to an expansion process

$$\Phi = \left(\lambda + \frac{2\mu}{3} \right) (\operatorname{div} \mathbf{V})^2 \geq 0 \tag{9.133}$$

and

$$\lambda + \frac{2\mu}{3} \geq 0 \tag{9.134}$$

The Stokes approximation consists of setting

$$3\lambda + 2\mu = 0 \tag{9.135}$$

The expansion of the fluid volume is not considered to be a source of dissipation. This assumption is approximatively true. For diluted monoatomic gases it is genuine.

In Chapter 2 on the constitutive laws for fluid materials we have written

$$T_{ik} = -p \, \delta_{ik} + A \, \delta_{ik} \operatorname{div} \mathbf{V} + 2B \left(D_{ik} - \frac{1}{3} \operatorname{div} \mathbf{V} \, \delta_{ik} \right) \tag{9.136}$$

The engineer preferentially uses the energy equation with respect to the center of gravity of a fluid particle in the form

$$\rho \frac{d\mathscr{E}}{dt} = -\frac{\partial q_i}{\partial x_i} - p \ \text{div} \ \mathbf{V} + \Phi \tag{9.137}$$

If we introduce Fourier's law,

$$q_i = \frac{\partial T}{\partial x_i} \tag{9.138}$$

We obtain

$$\boxed{\rho C_v \left\{ \frac{\partial T}{\partial t} + V_j \frac{\partial T}{\partial x_j} \right\} = k \frac{\partial^2 T}{\partial x_i \, \partial x_i} - p \ \text{div} \ \mathbf{V} + \Phi} \tag{9.139}$$

For low-speed flows, the amount of heat due to friction losses Φ is negligible. Moreover, if the dilatation of the fluid is also weak, the term $-p \ \text{div} \ \mathbf{V}$ can be dropped, and Eq. 9.139 exhibits a balance between convective and diffusive effects:

$$\rho C_v \left\{ \frac{\partial T}{\partial t} + V_j \frac{\partial T}{\partial x_j} \right\} = k \frac{\partial^2 T}{\partial x_i \, \partial x_i} \tag{9.140}$$

In Chapter 7, devoted to boundary layer theory, we introduced the energy equation in this form.

For high-speed flows, the problem is considerably modified. If the stream temperature is T_0, the temperature at the point A of a blunt body is T_A (Fig. 9.14). The two temperatures T_A and T_0 differ by ΔT, which is termed the *kinetic temperature*.

$$T_A - T_0 = \Delta C \tag{9.141}$$

At point B, the wall temperature is T_B.

$$T_B - T_0 = r \ \Delta C \tag{9.142}$$

where r is the recovery factor ($r \sim 0.8$).

Some order of magnitude can be given:

$$C_p (T_A - T_0) = \frac{V^2}{2} \tag{9.143}$$

if the speed $|\mathbf{V}|$ is equal to 300 m/s, we obtain

$$T_A - T_0 = \Delta C \simeq \frac{V^2}{2000} \simeq 15°C$$

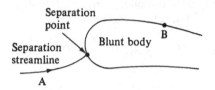

Fig. 9.13   Flow past a blunt body

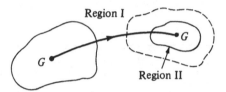

Fig. 9.14   Elementary motion of a fluid blob

If $|\mathbf{V}|$ is equal to 600 m/s, we find that

$$\Delta C = T_A - T_0 \simeq 180°C$$

and for $|\mathbf{V}|$, equal to 900 m/s,

$$\Delta C \simeq 405°C$$

At point B, the increase of T is linked to the role of the dissipation.

   Remark: We must reexamine the energy equation deduced from the law of mechanics,

$$\frac{dV_i}{dt} = \rho f_i + \frac{\partial}{\partial x_k}(T_{ki}) \tag{9.144}$$

which yields

$$\int_V V_i \frac{dV_i}{dt} \rho \, d\mathcal{V} = \int_V f_i \, V_i \, \rho \, d\mathcal{V} + \int_V V_i \frac{\partial}{\partial x_k}(T_{ki}) \, d\mathcal{V} \tag{9.145}$$

The work of the contact forces acting on the fluid inside the volume $\mathcal{V}$ is given by

$$\int_V \frac{\partial}{\partial x_k}(T_{ki} \, V_i) \, d\mathcal{V} = \int_S T_{ki} V_i \, n_k \, dS \tag{9.146}$$

This work includes all the forces acting on the volume of fluid with the exception of the volume forces **f.** Volume forces and surface forces contribute to accelerate the mass of fluid included in $\mathcal{V}$ except for an amount fixed by the dissipation function.

   On the other hand it is normal for the acceleration of the centers of mass not to be affected by the work of forces acting with respect to these centers of gravity. The following computation accounts for these remarks:

$$\frac{d}{dt} \int_v \frac{V_i^2}{2} \rho \, d\mathcal{V} = \int_v f_i V_i \rho \, d\mathcal{V} + \int_S T_i V_i \, dS - \int_v T_{ki} \frac{\partial V_i}{\partial x_k} d\mathcal{V} \qquad (9.147)$$

The surface integral $\int_s T_i V_i \, dS$ takes into account the two elementary motions represented in Fig. 9.15 to evaluate the work of the surface stresses:

$$W_{\text{global}} = W_{\text{I}} + W_{\text{II}}$$

$$\mathcal{P}_t = \mathcal{P}_{\text{global}} = \mathcal{P}_{\text{I}} + \mathcal{P}_{\text{II}}$$

where $\mathcal{P}$ is the rate of work. $\mathcal{P}_e$ is the corresponding term for the volume forces. Thus,

$$\frac{dC}{dt} = \mathcal{P}_e + \mathcal{P}_t - \int T_{ki} \frac{\partial V_i}{\partial x_k} d\mathcal{V} \qquad (9.148)$$

For a compressible fluid, the last term does contain the dissipation rate as we will see. We recall that

$$T_{ki}\left(\frac{\partial V_i}{\partial x_k}\right) = \frac{1}{2}\left\{ T_{ki} \frac{\partial V_i}{\partial x_k} + T_{ik} \frac{\partial V_k}{\partial x_i} \right\} = T_{ik} D_{ik} \qquad (9.149)$$

and

$$T_{ki} = -p\delta_{ik} + \lambda \operatorname{div} \mathbf{V} \, \delta_{ik} + 2\mu \, D_{ik} \qquad (9.150)$$

Therefore, we have

$$T_{ik} D_{ik} = -p \operatorname{div} \mathbf{V} + \lambda \, (\operatorname{div} \mathbf{V})^2 + 2\mu \, D_{ik} D_{ik} \qquad (9.151)$$

with

$$D_{ii} = \operatorname{div} \mathbf{V} = \rho \frac{d(1/\rho)}{dt} \qquad \text{continuity equation} \qquad (9.152)$$

The evolution of the kinetic energies of all the fluid particles obeys the next equation:

$$\frac{dC}{dt} = \mathcal{P}_e + \mathcal{P}_t - \left\{ -\int_v p \operatorname{div} \mathbf{V} \, d\mathcal{V} \right\} - \int_v \Phi \, d\mathcal{V} \qquad (9.153)$$

$$\underbrace{\qquad\qquad}_{\mathcal{P}_{\text{I}} + \mathcal{P}_{\text{II}}} \quad - \quad \underbrace{\qquad\qquad}_{\mathcal{P}_{\text{II}}}$$

$$\boxed{\frac{dC}{dt} = \mathcal{P}_e + \mathcal{P}_{\text{I}} - \int_v \Phi \, d\mathcal{V}} \qquad (9.154)$$

dissipation rate

The acceleration of the centers of gravity of the fluid particles is hampered by the intrinsic nature of the fluid medium through the dissipation rate. The part of the forces that play a role in this motion are $\mathscr{P}_e$ and $\mathscr{P}_I$.

Appendixes

A   The Concept of "Objective Time Derivatives"

The mechanisms of continuous media are concerned with the motion and deformation of bodies. A reference frame is necessary to describe the phenomena occurring in the medium under consideration. Because the fundamental measurable quantities of classical kinematics are distance and time intervals, any change of reference frame must preserve distance and time intervals. In other words, any mapping of space-time into itself that would lead to the stretching or contraction of space or time is not appropriate in this case. According to this basic restriction, the most general change of frame is given by

$$\mathbf{X}^*(t) = \mathbf{b}(t) + \mathbf{Q}(t)\mathbf{x}(t) \tag{A.1}$$

or in an equivalent form,

$$x_k^*(t) = b_k(t) + Q_{kl}(t)x_l(t) \tag{A.2}$$

$\mathbf{x}(t)$ denotes some element of the continuous medium at time t, in the reference frame $\mathfrak{R}$, $\mathbf{x}^*(t)$ denotes the same element in the reference frame $\mathfrak{R}^*$, $\mathbf{b}(t)$ corresponds to translation, and $\mathbf{Q}(t)$, which is a time-dependent orthogonal tensor, corresponds to rotation.

A.2 THE CONCEPT OF FRAME INDIFFERENCE (OR MATERIAL OBJECTIVITY)

A.2.a *Definition*

A physical quantity is said to be frame independent (or to obey the principle of material objectivity) if it remains invariant under all changes of reference frame $\mathfrak{R} \to \mathfrak{R}^*$. If this quantity is

    a. a scalar, we have $S^* = S$
    b. a vector, then $V_k^* = Q_{kl}V_l$
    c. a second-order tensor, then $T_{kl}^* = Q_{km}T_{mn}Q_{ln}$

Let us multiply some objective vector, say V_k by a second-order tensor T_{kl} (not necessarily objective a priori). If the resulting vector W_k is

275

objective, T_{kl} is also shown to be objective, which means that T_{kl} satisfies a relation of the third form (c). This statement is easy to prove. We first write

$$W_k = T_{kl}V_k \qquad (A.3)$$

V_k, being assumed to be an objective tensor equation (b), is verified:

$$V_k^* = Q_{kl}V_l \qquad (A.4)$$

Moreover, W_k is an objective tensor, which gives

$$W_k^* = Q_{kl}W_l \qquad (A.5)$$

In the reference frame $\mathfrak{R}^*$, Eq. A.3 has the following form:

$$W_k^* = T_{kl}^*V_l^* \qquad (A.6)$$

Using Eqs. A.4 and A.5 to substitute V_l^* and W_k^*, respectively, we can rewrite Eq. A.3 as

$$Q_{kl}W_l = T_{kl}^*Q_{lm}V_m \qquad (A.7)$$

From Eq. A.3, we substitute W_l into Eq. A.7 to obtain

$$Q_{kl}T_{lm}V_m = T_{kl}^*Q_{lm}V_m \qquad (A.8)$$

Equation A.8 holds for any vector V_m which implies that

$$Q_{kl}T_{lm} = T_{kl}^*Q_{lm} \qquad (A.9)$$

Multiplying both sides of Eq. A.9 by the transposed tensor, and after elementary manipulations on indices, we have

$$Q_{km}T_{mn}Q_{ln} = T_{kl}^* \qquad (A.10)$$

which is the condition (c) for T to be an objective tensor.

A.2.b   *Examples*

Depending on their physical meanings, some quantities (vectors or tensors) are objective and others are not.

Velocity: Differentiation Eq. A.2 with respect to time yields

$$\dot{x}_k^*(t) = \dot{b}_k(t) + \dot{Q}_{kl}(t)x_l(t) + Q_{kl}(t)\dot{x}_l(t) \qquad (A.11)$$

Rearranging of the terms and introducing the velocity vector $v_k(t)$ gives

$$v_k^*(t) = Q_{kl}(t)\dot{x}_l(t) + \dot{b}_k(t) + \dot{Q}_{kl}(t)x_l(t) \qquad (A.12)$$

Due to the additional terms,

$$\dot{b}_k(t) + \dot{Q}_{kl}(t)x(t) \qquad (A.13)$$

$v_k^*(t)$ cannot be reduced to $Q_{kl}(t)\dot{x}_l(t)$.

Rate-of-strain tensor: In the two reference frames $\mathscr{R}$ and $\mathscr{R}^*$, the strain tensor is, respectively, represented as follows:

$$D_{kl} = \frac{1}{2}\left[\frac{\partial v_k}{\partial x_l} + \frac{\partial v_l}{\partial x_k}\right] \tag{A.14}$$

and

$$D_{kl}^* = \frac{1}{2}\left[\frac{\partial v_k^*}{\partial x_l^*} + \frac{\partial v_l^*}{\partial x_k^*}\right] \tag{A.15}$$

Let us write $\partial v_k^*/\partial x_l^*$ in a different way by using Eq. A.12:

$$\frac{\partial v_k^*}{\partial x_l^*} = Q_{km}\frac{\partial v_m}{\partial x_l^*} + \dot{Q}_{km}\frac{\partial x_m}{\partial x_l^*} \tag{A.16}$$

$$\frac{\partial v_k^*}{\partial x_l^*} = Q_{km}\frac{\partial v_m}{\partial x_n}\frac{\partial x_n}{\partial x_l^*} + \dot{Q}_{lm}\frac{\partial x_m}{\partial x_l^*} \tag{A.17}$$

Let us express $\partial x_m/\partial x_l^*$ by using Eq. (A.2):

$$x_k^* = b_k + Q_{kl}x_l$$

If both sides of this equation are multiplied by Q_{km}, we readily obtain

$$Q_{km}(x_k^* - b_k) = Q_{km}Q_{kl}x_l = \delta_{ml}x_l \tag{A.18}$$

This equation can be rewritten as

$$Q_{kl}(x_k^* - b_k) = x_l \tag{A.19}$$

from which we infer

$$\frac{\partial x_l}{\partial x_k^*} = Q_{kl} \tag{A.20}$$

Equation A.17 can be rearranged:

$$\frac{\partial v_k^*}{\partial x_l^*} = Q_{km}\frac{\partial v_m}{\partial x_n}Q_{ln} + \dot{Q}_{km}Q_{lm} \tag{A.21}$$

In the same way we find that

$$\frac{\partial v_l^*}{\partial x_k^*} = Q_{lm}\frac{\partial v_m}{\partial x_n}Q_{kn} + \dot{Q}_{lm}Q_{km} \tag{A.22}$$

An equivalent form is written, after elementary manipulations, as

$$\frac{\partial v_l^*}{\partial x_k^*} = Q_{ln}\frac{\partial v_n}{\partial x_m}Q_{km} + \dot{Q}_{lm}Q_{km} \tag{A.23}$$

The expression of D_{kl}^* is obtained by adding Eqs. A.21 and A.23:

$$2D_{kl}^* = Q_{km}\left[\frac{\partial v_m}{\partial x_n} + \frac{\partial v_n}{\partial x_m}\right]Q_{ln} + \frac{d}{dt}(Q_{km}Q_{lm}) \qquad (A.24)$$

which reduces to

$$D_{kl}^* = Q_{km}D_{mn}Q_{ln} \qquad (A.25)$$

since we have

$$\frac{d}{dt}(Q_{km}Q_{lm}) = 0 = \frac{d}{dt}(\delta_{kl}) \qquad (A.26)$$

Hence, D_{kl} is objective because D_{kl} and D_{kl}^* are related by means of a relation of the third type (c).

Vorticity: The vorticity tensor is defined as follows:

$$\omega_{kl} = \frac{1}{2}\left[\frac{\partial u_k}{\partial x_l} - \frac{\partial u_l}{\partial x_k}\right] \qquad (A.27)$$

Using the same calculation as for the rate-of-strain tensor, we obtain

$$\omega_{kl}^* = Q_{km}\omega_{mn}Q_{ln} + \dot{Q}_{km}Q_{lm} \qquad (A.28)$$

The vorticity tensor is not objective ($\omega_{kl}^* \neq Q_{km}\omega_{mn}Q_{ln}$) because of the additional term that represents the angular velocity of the frame $\mathscr{R}^*$ with respect to $\mathscr{R}$.

A.3   THE CONCEPT OF AN OBJECTIVE TIME DERIVATIVE

A.3.a   *Definition*

The foregoing calculations display the existence of quantities that are objective whereas others are not. The reason depends on their physical nature. If the rheological properties of a continuous medium depend on the local geometry, such as the distance, the properties of the medium have to be objective. This is the case for the rate-of-deformation tensor or the stress tensor. To refine the descriptions of these quantities, we introduce the concept of time derivatives. In so doing, the operator called *time derivative* must not alter the initial property of objectivity, which is a straightforward consequence of the physical nature of the quantity under consideration. In other words, an objective quantity must remain so after we take its derivative with respect to time. The time derivatives we consider must meet this property. This assertion is equivalent to the following simple formulation.

Let T_{kl} be a second-order objective tensor. Its time derivative, denoted by $\mathscr{D}T_{kl}/\mathscr{D}t$ has to satisfy a relation of the third class (c). (In fact, this time derivative is an operator.)

$$\frac{\mathfrak{D}T_{kl}^*}{\mathfrak{D}t} = Q_{km}\frac{\mathfrak{D}T_{mm}}{\mathfrak{D}t}Q_{ln} \tag{A.29}$$

The second-order tensor T_{kl} is assumed to obey relation (c), which means

$$T_{kl}^* = Q_{km}T_{mn}Q_{ln} \tag{A.30}$$

A.3.b   Material derivatives

To take into account memory effects, we could choose material derivatives as time derivatives. In so doing, we could introduce the history of the rate-of-strain tensor. Classical advective time derivative leads to

$$\frac{DT_{kl}^*}{Dt} = Q_{kn}\frac{DT_{mn}}{Dt}Q_{ln} + \frac{DQ_{kn}}{Dt}T_{mn}Q_{ln} + Q_{km}T_{mn}\frac{DQ_{ln}}{Dt} \tag{A.31}$$

Clearly this equation does not reduce to Eq. A.30 by reason of two additional terms involving the advective time derivative of the tensor **Q**. Accordingly, this time derivative is not objective; it is useless for the description of the rheological behavior of contiumm media as well as for the derivation of constitutive laws.

A.4   OBJECTIVE TIME DERIVATIVE OPERATORS

A.4.a   General remarks

Starting from Eq. A.31, we now introduce a time derivative that is objective for any second-order term (i.e., any second-order tensor verifying Eq. A.30). For this purpose, various expressions involving ω_{kl}, ω_{kl}^*, $\partial u_k/\partial x_l$, and $\partial u_k^* \partial x_l^*$ are used. We have to substitute these new expressions for DQ_{kl}/Dt in Eq. A.31. More than a single expression can be obtained, but only one of them is presented hereafter.

A.4.b   Example

Multiplying both sides of Eq. A.28 by Q_{kl} and solving with respect to $\dot{Q}_{kl}$, we have

$$\dot{Q}_{kl} = \omega_{km}^* Q_{ml} - Q_{km}\omega_{ml} \tag{A.32}$$

Equation A.31 becomes

$$\frac{DT_{kl}^*}{Dt} = Q_{km}\frac{DT_{mn}}{Dt} + \omega_{kr}^* Q_{rm}T_{mn}Q_{ln} - Q_{kr}\omega_{rm}T_{mn}Q_{ln}$$

$$+ Q_{kn}T_{mn}\omega_{lr}^* Q_{rn} - Q_{km}T_{mn}Q_{lr}\omega_{rn} \tag{A.33}$$

The second and the fourth terms on the right-hand side of Eq. A.33 can be rewritten in a simplified form by noting that $Q_{rm}T_{mn}Q_{ln}$ and $Q_{km}T_{mn}Q_{rn}$ are equal to T_{rl}^* and T_{kr}^*, respectively:

$$\frac{DT_{kl}^*}{Dt} - \omega_{kr}^* T_{rl}^* - \omega_{lr}^* T_{kr}$$

$$= Q_{km}\frac{DT_{mr}}{Dt}Q_{ln} - Q_{kr}\omega_{rm}T_{mn}Q_{ln} - Q_{km}T_{mn}Q_{lr}\omega_{rn} \qquad (A.34)$$

Equation A.34 is rearranged after elementary manipulation of indices to yield

$$\frac{DT_{kl}^*}{Dt} - T_{ml}^*\omega_{km}^* + T_{km}^*\omega_{ml}^*$$

$$= Q_{km}Q_{ln}\left[\frac{DT_{mn}}{Dt} - \omega_{mr}T_{rn} + T_{mr}\omega_{rn}\right] \qquad (A.35)$$

We now set

$$\frac{\mathcal{D}T_{kl}}{\mathcal{D}t} = \frac{D}{Dt}(T_{kl}) - \omega_{km}T_{ml} + T_{km}\omega_{ml} \qquad (A.36)$$

Equation A.35 appears to be equivalent to

$$\frac{\mathcal{D}T_{kl}}{\mathcal{D}t} = Q_{km}\frac{\mathcal{D}}{\mathcal{D}t}(T_{mn})Q_{ln} \qquad (A.37)$$

and, accordingly, the operator $\mathcal{D}/\mathcal{D}t$ is an objective operator (Eq. A.29). This objective time derivative operator is called the *Jaumann–Noll derivative*. As previously mentioned, other types of derivatives can be inferred in a similar way. Let us mention, for instance, the *Oldroyd derivative*

$$\frac{D}{Dt}(T_{kl}) - T_{ml}\frac{\partial u_k}{\partial x_m} - T_{km}\frac{\partial u_l}{\partial x_m} \qquad (A.38)$$

and the *codeformational derivative*

$$\frac{D}{Dt}(T_{kl}) + T_{km}\frac{\partial u_m}{\partial x_l} + T_{ml}\frac{\partial u_m}{\partial x_k} \qquad (A.39)$$

The choice of the most convenient derivative is dictated by the structure of the medium.

B   Partial Differential Equations of Mathematical Physics

B.1   INTRODUCTION

Physical phenomena generally depend on more than one independent variable. Accordingly, they obey partial differential equations. Reference can be made to some typical cases encountered in various chapters on fluid mechanics.

If we assume a fluid to be both inviscid and incompressible, we can introduce a velocity potential. That obeys the Laplace equation:

$$\frac{\partial^2 \Phi}{\partial x^2} + \frac{\partial^2 \Phi}{\partial y^2} = 0 \tag{B.1}$$

The typical behavior of this equation can be stated in several ways. If, for instance, we introduce a disturbance in the boundary conditions $(\partial \Phi / \partial n|_s = 0)$ the whole flow is disturbed and the disturbance becomes smaller as the observation point goes farther and farther away from a bump (Fig. B.1).

We could deal with partial differential equations that look like wave equations, but we prefer to analyze a similar problem starting from the Steichen equation, which deals with compressible flows,

$$\frac{\partial^2 \Phi}{\partial x^2} \left(1 - \frac{u^2}{a^2} \right) - 2 \frac{\partial^2 \Phi}{\partial x \, \partial y} \frac{uv}{a^2} + \frac{\partial^2 \Phi}{\partial y^2} \tag{B.2}$$

where a denotes the speed of sound.

If the flow is supersonic, that is, if $u^2 + v^2 > a^2$, the disturbance introduced by the small bump is located in physical space (see Fig. B.2). Besides, the disturbance is not a decreasing function of the distance from the bump. This behavior is characteristic of supersonic flows. If the speed of sound approaches infinity, the Steichen equation becomes a Laplace equation. The type of equation depends on the Mach number M, where

$$M = \frac{u^2 + v^2}{a^2} \qquad \begin{cases} M < 1, \text{ subsonic flows} \\ M > 1, \text{ supersonic flows} \end{cases}$$

With heat exchanges, we encounter another situation. A step change

281

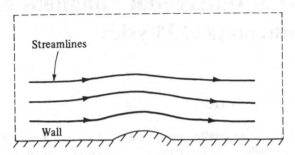

Fig. B.1   Streamlines are disturbed everywhere by a bump in the wall

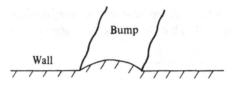

Fig. B.2   Two-dimensional bump

of temperature at the wall instantaneously disturbs the whole thermal field, and the disturbance is a decreasing function of the distance from the location of the step change. From these examples we expect the problem of determining the solution of a partial differential equation to be defined by typical boundary conditions.

For an elliptic equation, the Dirichlet conditions are required at the boundary. The following example provides physical guidance. Boundary conditions are given along the wall, which is assumed to extend to infinity, and along the contour $ABCD$, which appears to be an extension of the wall line. The velocity distribution along the bump is not determined by local conditions; the whole flow is involved. In fact, the combined following conditions are required:

$$\text{for AD,} \quad \frac{\partial \Phi}{\partial n} = 0 \qquad \text{for BC,} \quad \frac{\partial \Phi}{\partial n} = 0$$

$$\text{for AB,} \quad \frac{\partial \Phi}{\partial n} = U_0 \qquad \text{for CB,} \quad \frac{\partial \Phi}{\partial n} = 0$$

$$S \quad \overset{\Delta \Phi = 0}{\bigcirc} \leftarrow \Phi$$

A unique regular solution of Φ is determined inside the domain $\mathcal{V}$, limited by the surfaces (Dirichlet problem) for given value of Φ on S. Internal or external problems can then be solved.

If the derivative of Φ is prescribed at any point of S, the function Φ being unknown, (see figure) the solution is still defined except for an additional constant (Neumann problem). For the physical example, Dirichlet and Neumann conditions are mixed.

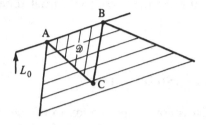

For a hyperbolic equation, the Cauchy conditions are required along the curve L_0, which means that Φ and the first derivatives of Φ are known along L_0 (see figure). The solution is completely defined in a domain $\mathcal{D}$ limited in extent. AC and BC are two characteristic lines from A and B, respectively. $\mathcal{D}$ is the *domain of dependence*. Cauchy's data are capable of defining a solution of the problem in this area. These same data interfere with the solution of the problem in an infinite sector, which is called the *domain of influence*. The second physical example also provides some guidance; the disturbance introduced by the bump is also limited to a stripe. On the bump, the magnitude of the velocity vector is linked to the local deviation imposed by the curvature of the wall, so Cauchy conditions are effectively introduced (Φ and $U_S = \partial\Phi/\partial S$). This local dependence characterizes supersonic flows. Although the presence of an opposite wall locally alters the velocity distribution in region I (subsonic flow) this additional boundary condition does not disturb the velocity distribution in region II (supersonic flow); see figure.

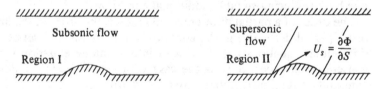

For a parabolic equation such as that for heat conduction, the problem is determined by the amplitude of the step change in temperature introduced at the wall at time $t = 0$. The additional conditions on the derivatives are required to solve this problem, as it can be shown by using a finite element scheme.

Three types of second-order schemes can be identified, each of which corresponds to a typical physical problem. Through these examples, an intuitive classification has been introduced. This classification is actually introduced starting from a unique second-order equation, and the theory is also restricted to first- and second-order partial differential equations.

In boundary layer theory, reference is made to a set of two or three partial differential equations; so the problem appears to be set in a somewhat unusual way. The analysis in this chapter should make this approach more understandable, but by no means should this be considered more than an introduction to the theory of partial differential equations.

B.2   BASIC CONCEPTS AND DEFINITION

A partial differential equation contains, in addition to the dependent and independent variables, one or more derivatives of the dependent variables. In general, it is written in the form

$$f\left(x, y, \ldots, \Phi, \frac{\partial \Phi}{\partial x}, \ldots, \frac{\partial^2 \Phi}{\partial x^2}, \frac{\partial^2 \Phi}{\partial x \partial y}, \ldots\right) = 0 \qquad (B.3)$$

and involves several independent variables $x, y, \ldots$, an unknown function Φ of these variables, and the partial derivatives $\partial \Phi/\partial x$, $\partial \Phi/\partial y, \ldots$, $\partial^2 \Phi/\partial x^2$, $\partial^2 \Phi/\partial y^2$, $\ldots$ of the functions. For the sake of brevity, such an equation is often written as

$$f(x, y, \Phi, \Phi_x, \ldots, \Phi_{xx}, \Phi_{xy}, \ldots) = 0 \qquad (B.4)$$

It is considered in a suitable domain $\mathscr{D}$ of the n-dimensional space R^n in direct relation with the n independent variables $x, y, \ldots$. We seek functions $\Phi(x, y, \ldots)$ that satisfy the previous equation in $\mathscr{D}$. Such functions if they exist, are called *solutions* of the partial differential equation. From these many possible solutions, we select a particular one by introducing suitable additional conditions.

The order of a partial differential equation is the order of the highest-order partial derivative in the equation. In many cases, a set of partial differential equations is transformed into a unique equation having a consequential order. Such is the case with incompressible fluids. The continuity equation is initially associated with the condition that curl $V = 0$:

$$\left. \begin{array}{l} \dfrac{\partial u}{\partial x} + \dfrac{\partial v}{\partial y} = 0 \\[2mm] \dfrac{\partial v}{\partial x} - \dfrac{\partial u}{\partial y} = 0 \Rightarrow u = \mathbf{grad}\ \Phi \end{array} \right\} \Delta \Phi = 0 \qquad (B.5)$$

Similar steps lead to the Steichen equation, which supposes a barotropic flow $p = p(\rho)$ and introduces the speed of sound a:

$$u \frac{\partial u}{\partial x} + v \frac{\partial u}{\partial y} = -\frac{1}{\rho} \frac{\partial p}{\partial x} = -\frac{1}{\rho} \frac{\partial p}{\partial \rho} \frac{\partial \rho}{\partial x} = -\frac{a^2}{\rho} \frac{\partial \rho}{\partial x} \qquad (B.6)$$

$$u\,\frac{\partial v}{\partial x} + v\,\frac{\partial v}{\partial y} = -\frac{1}{\rho}\,\frac{\partial p}{\partial y} = -\frac{1}{\rho}\,\frac{\partial p}{\partial y} = -\frac{a^2}{\rho}\,\frac{\partial \rho}{\partial y} \tag{B.7}$$

$$\frac{\partial(\rho u)}{\partial x} + \frac{\partial(\rho v)}{\partial y} = 0 \tag{B.8}$$

After some rearrangements, we obtain the Steichen equation previously introduced; a similar situation is encountered for wave equations. However, note that it may be useless to transform a set of partial differential equations into a unique equation of higher order. The boundary layer equations will be treated as a set of partial differential equations. We shall accept some restraints; for instance, the number of independent variables will be limited to two, the number of functions of those independent variables being unlimited. In doing so, the problem of a steady two-dimensional boundary layer becomes tractable.

A partial differential equation is said to be linear if it is linear with respect to the unknown function and all its derivatives with coefficients depending only on the independent variables. The equation

$$y\Phi_{xx} + 2xy\Phi_{yy} + yu = 1 \tag{B.9}$$

is a nonhomogeneous second-order linear partial differential equation, whereas

$$\Phi_{yy}\Phi_{xx} + x\Phi\Phi_y = 1 \tag{B.10}$$

is nonlinear.

The most general form of second-order linear equation involving n independent variables has the form

$$\sum_{i,j=1}^{n} A_{ij}\Phi_{x_i x_j} + \sum_{i=1}^{n} B_i\Phi_{x_i} + F\Phi = G \tag{B.11}$$

where we can assume that $A_{ij} = A_{ji}$. Moreover, we assume B_i, F, and G to be functions of the n independent variables x_i. For $G = 0$, the equation becomes homogeneous; otherwise, it is nonhomogeneous.

A partial differential equation is said to be quasilinear if it is linear in the highest order of the unknown function.

$$\Phi_x\Phi_{xx} + x\Phi\Phi_y = 0 \tag{B.12}$$

is a second-order quasilinear equation. The Steichen equation is of this type. Quasilinear partial differential equations can be treated by Cauchy's method.

Differential equations can be considered as differential operators. An operator is a mathematical operation that, when applied to a function, gives another function. We can define a laplacian operator L as

$$L = \frac{\partial^2}{\partial x^2} + \frac{\partial^2}{\partial y^2} \tag{B.13}$$

or a d'alembertian operator D as

$$D = \frac{\partial^2}{\partial t^2} - a^2 \frac{\partial^2}{\partial x_i \, \partial x_i} \qquad (B.14)$$

and then write

$$L[\Phi] = \frac{\partial^2 \Phi}{\partial x^2} + \frac{\partial^2 \Phi}{\partial y^2} \qquad (B.15)$$

and

$$D[\Phi] = \frac{\partial^2 \Phi}{\partial t^2} - a^2 \frac{\partial^2 \Phi}{\partial x_i \, \partial x_i} \qquad (B.16)$$

Linear differential operators are associated with linear operators that satisfy some classical properties:

1. A constant C can be moved outside the operator:

$$L[C\Phi] = CL[\Phi] \qquad (B.17)$$

2. The operator operating on the sum of two functions gives the sum of the operator operating on the individual functions:

$$L[\Phi_1 + \Phi_2] = L[\Phi_1] + L[\Phi_2] \qquad (B.18)$$

Combining the two previous properties, we have

$$L[C_1\Phi_1 + C_2\Phi_2] = C_1 L[\Phi_1] + C_2 L[\Phi_2] \qquad (B.19)$$

Linear differential operators satisfy the following properties:

$L + M = M + L$ (commutative)
$(L + M) + N = L + (M + N)$ (associative)
$(LM)N = L(MN)$ (associative)
$L(M + N) = LM + LN$ (distributive)

Linear operators are not commutative ($LM \neq ML$) except for linear differential operators with constant coefficients.

B.3 PRELIMINARY REMARKS

The treatment of partial differential equations may be introduced in several ways that finally emphasize the role of lines that are capable of generating the surface solution. Even though the method developed here can be extended to more general cases, we limit our approach to two independent variables so that the solutions can be represented by one or several surfaces defined in a three-dimensional space. $\Phi(x, y)$ is called the integral surface of the problem.

The Cauchy problem is the most convenient way to treat partial differential equations. The initial step is similar to that introduced for treating ordinary differential equations by the initial value problem. First, this method consists of finding the solution of a given differential equation with an appropriate number of initial conditions prescribed at an initial point. For example, the second-order ordinary differential equation

$$\frac{d^2y}{dt^2} = f\left(x, y, \frac{dy}{dx}\right) \tag{B.20}$$

and the initial conditions

$$y(x_0) = \alpha \tag{B.21}$$

$$\frac{dy}{dt}(x_0) = \beta \tag{B.22}$$

constitute an initial value problem. A solution can usually be introduced in the form of a Taylor power series. Another step consists in demonstrating that the domain of convergence of this series is different from zero.

For second-order partial differential equation of the hyperbolic type, it is possible to show that an integral surface solution exists if the Cauchy data are prescribed. The integral surface is prescribed to pass through $\mathscr{L}$ (which is a line in three-dimensional space) and to admit along $\mathscr{L}$ first derivatives that are equally prescribed. If $\mathscr{L}$ is not a characteristic line, the Cauchy problem has a solution in a more or less restricted area of the space. In the case under consideration, two characteristic families exist that are capable of generating the integral surface. The differential forms in connection with these characteristic curves generally degenerate into a system of ordinary differential equations. Accordingly, the initial problem is associated with another that is more tractable. The concept of characteristic lines drawn on integral surfaces is fundamental in this kind of problem. It is introduced by solving the problem of Cauchy.

Another approach consists in introducing a local system of coordinates to reduce the initial partial differential equation to its canonical form. In fact, this system is linked to the characteristic lines at the point.

A heuristic approach of the characteristic lines can be introduced by dealing with a first-order partial differential equation. It emphasizes the close connection existing between the role of typical curves and the search for an integral surface, without referring to refined methods. It can be used as an introduction.

The first-order equation

$$\frac{\partial \Phi}{\partial x} + 2x \frac{\partial \Phi}{\partial y} = y \tag{B.23}$$

is associated with the condition

$$\Phi(0, y) = 1 + y^2 \quad \text{for } 1 < y < 2 \qquad \text{(B.24)}$$

If we assume the existence of a curve in the xy plane, such as $y = y(x)$, we can write along this curve

$$\frac{d}{dx} \Phi = \frac{\partial \Phi}{\partial x} + \frac{dy}{dx} \frac{\partial \Phi}{\partial y} \qquad \text{(B.25)}$$

By comparison with the left-hand side of the equation under consideration, we find that

$$\frac{dy}{dx} = 2x \qquad \text{(B.26)}$$

that is, $y = x^2 + C_1$ where C_1 is a constant.

Thus, we can interpret Eq. B.23 as a statement that along any surface $y = x^2 + C_1$,

$$\frac{d}{dx} \Phi(x, y(x)) = y = x^2 + C_1 \qquad \text{(B.27)}$$

Thus, along such a curve,

$$\Phi = \frac{1}{3} x^3 + C_1 x + C_2 \qquad \text{(B.28)}$$

where C_2 represents the value of Φ for $x = 0$. The parabolas $y = x^2 + C_1$ are drawn in Fig. B.3. Each curve is characterized by a value of C_1.

Along any one of these curves, Eq. B.28 is satisfied. If we know Φ at some point on the curve, we can use Eq. B.28 to determine C_2; thus, Φ can be known at all points on that curve. The boundary condition Eq. B.24 specifies $\Phi(0, y)$ for $1 < y < 2$. That enables us to determine $\Phi(x, y)$ at any point x, y that can be linked to the segment $(1, 2)$ of the y axis by means of one of the parabolas corresponding to the shaded region of Fig. B.3.

Let us call (x_0, y_0) a point at which we have to determine Φ. The parabola through (x_0, y_0) obeys the equation

$$y = x^2 + (y_0 - x_0^2) \Rightarrow C_1 = (y_0 - x_0^2) \qquad \text{(B.29)}$$

and this parabola cuts the y axis at the point $y = y_0 - x_0^2$. At this point, the boundary conditions Eq. B.24 require that

$$\Phi(0, y) = 1 + y^2 = 1 + (y_0 - x_0^2)^2 = C_2 \qquad \text{(B.30)}$$

The value of Φ at any other point of the parabola (Eq. B.29) is now given by Eq. B.28:

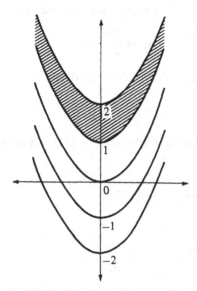

Fig. B.3   Characteristics

$$\Phi = \frac{1}{3} x^3 + (y_0 - x_0^2)x + \{1 + (y_0 - x_0^2)^2\} \qquad \text{(B.31)}$$

In particular, because the parabola passes through the point (x_0, y_0), we have

$$u(x_0, y_0) = \frac{1}{3} x_0^3 + (y_0 - x_0^2)x_0 + \{1 + (y_0 - x_0^2)^2\} \qquad \text{(B.32)}$$

But (x_0, y_0) is any point in the shaded region. Thus, by replacing x_0 by x and y_0 by y, we finally have the result at any point (x, y) in this region:

$$u(x, y) = x^4 - \frac{2}{3} x^3 - 2x^2 y + xy + y^2 + 1 \qquad \text{(B.33)}$$

It can be verified by direct substitution that Eq. B.23 associated with the boundary condition Eq. B.24 is satisfied.

This procedure has uniquely determined Φ throughout the shaded region, but no information has been collected outside the region. This example emphasizes the role of the parabolas in the treatment of partial differential equations. In fact, these parabolas are the characteristic system corresponding to the initial equation.

We have represented the method of characteristics in a very heuristic way. For first-order partial differential equations, the introduction of the characteristic system is generally obtained starting from the equation

$$f(x, y) \frac{\partial \Phi}{\partial x} + g(x, y) \frac{\partial \Phi}{\partial y} = 0 \qquad (B.34)$$

and from the associated ordinary differential equation

$$\frac{dx}{f(x, y)} = \frac{dy}{g(x, y)} = \alpha \qquad (B.35)$$

Let us suppose $\Phi(x, y) = C$ to be a first integral of the system, which leads to

$$\frac{\partial \Phi}{\partial x} dx + \frac{\partial \Phi}{\partial y} dy = 0 \qquad (B.36)$$

or

$$f(x, y) \frac{\partial \Phi}{\partial x} + g(x, y) \frac{\partial \Phi}{\partial y} = 0 \qquad (B.37)$$

This equation corresponds to the initial partial differential equation. Accordingly, solutions of this equation are the first integral of the associated system, which corresponds to characteristic lines drawn in the xy plane.

In the previous example, the associated ordinary differential equation would be written

$$\frac{dx}{1} = \frac{dy}{2x} \Rightarrow \frac{dy}{dx} = 2x \qquad (B.38)$$

In the case of a first-order partial differential equation, the characteristic curves are necessary and real.

B.4 THE CAUCHY PROBLEM FOR A SECOND-ORDER PARTIAL DIFFERENTIAL EQUATION

We limit the theory to the case of a second-order partial differential equation, called a *Monge–Ampere equation*. It can be written in the form

$$H \frac{\partial^2 \Phi}{\partial x^2} + 2K \frac{\partial^2 \Phi}{\partial x \partial y} + L \frac{\partial^2 \Phi}{\partial y^2} + M = 0 \qquad (B.39)$$

in which H, K, L, and M are generally functions of x, y, Φ, $\partial\Phi/\partial x$, and $\partial\Phi/\partial y$. Accordingly, it is a quasilinear equation. The following notations are used

$$r = \frac{\partial^2 \Phi}{\partial x^2} \quad s = \frac{\partial^2 \Phi}{\partial x \partial y} \quad t = \frac{\partial^2 \Phi}{\partial y^2} \quad p = \frac{\partial \Phi}{\partial x} \quad q = \frac{\partial \Phi}{\partial y} \qquad \text{(B.40)}$$

and the Monge–Ampere equation is written in the form

$$Hr + 2Ks + Lt + M = 0 \qquad \text{(B.41)}$$

As previously mentioned, the solution can be introduced in the form of an integral surface that is prescribed to pass through $\mathscr{L}$ and to admit along $\mathscr{L}$ the first derivatives which are equally prescribed. The curve $\mathscr{L}$ can be defined in parametric form, that is,

$$x = x(\sigma) \qquad y = y(\sigma) \qquad z = z(\sigma) = \Phi(\sigma)$$

Moreover, the direction cosines of the normals to the planes tangent to the surface along $\mathscr{L}$ are also prescribed; these are

$$p = p(\sigma) \qquad q = q(\sigma) \qquad -1$$

The functions $x, y, z = \Phi, p$, and q are assumed to be regular in the domain under consideration.

In fact, not all of the data are completely independent of each other; the orientations of the planes tangent to the surface are closely related to the orientations of the tangents to the curve $\mathscr{L}$. Thus, we have

$$\frac{dz}{d\sigma} = \frac{d\Phi}{d\sigma} = \frac{\partial \Phi}{\partial x}\frac{\partial x}{\partial \sigma} + \frac{\partial \Phi}{\partial y}\frac{\partial y}{\partial \sigma} \qquad \text{(B.42)}$$

$$= p\frac{dx}{d\sigma} + q\frac{dy}{d\sigma} \qquad \text{(B.43)}$$

The choice of Cauchy's data are accordingly restricted.

Consider a point (x, y) of the curve $\mathscr{L}$. The coefficients of the Monge–Ampere equation and the unknown function itself are assumed to be analytic, so the solution can be represented in the form of a Taylor series:

$$\Phi(x, y) = \sum_{n=0}^{\infty} \sum_{k=0}^{n} \frac{1}{k!(n-k)!} \frac{\partial^n \Phi_0}{\partial x_0^k \partial y_0^{n-k}} (x - x_0)^k (y - y_0)^{n-k} \qquad \text{(B.44)}$$

whose first terms are written as

$$\Phi(x, y) = \Phi_0(x_0, y_0) + (x - x_0)p_0 + (y - y_0)q_0$$

$$+ \frac{1}{2}[(x - x_0^2)r_0 + 2(x - x_0)(y - y_0)s_0$$

$$+ (y - y_0)^2 t_0] + \cdots \tag{B.45}$$

The Cauchy data can be shown to be sufficient to determine all the higher-order derivatives. Moreover, the Taylor series, which represents the solution in the vicinity of point (x_0, y_0), can also be shown to converge in a more or less extended domain about (x_0, y_0).

A point (x_0, y_0, z_0) of three-dimensional space that contains the integral surface corresponds to point (x_0, y_0) of the xy plane. All the quantities,

$$H_0 = H(x_0, y_0, z_0, p_0, q_0) \tag{B.46}$$

$$K_0 = K(x_0, y_0, z_0, p_0, q_0) \tag{B.47}$$

$$M_0 = M(x_0, y_0, z_0, p_0, q_0) \tag{B.48}$$

are known. We have to show that the derivatives of higher order must also be determined. In fact, we find that

$$\frac{dp}{d\sigma} = \frac{\partial p}{\partial x}\frac{dx}{d\sigma} + \frac{\partial p}{\partial y}\frac{dy}{d\sigma} = r\frac{dx}{d\sigma} + s\frac{dy}{d\sigma} \tag{B.49}$$

$$\frac{dq}{d\sigma} = \frac{\partial q}{\partial x}\frac{dx}{d\sigma} + \frac{\partial q}{\partial y}\frac{dy}{d\sigma} = s\frac{dx}{d\sigma} + t\frac{dy}{d\sigma} \tag{B.50}$$

At the point (x_0, y_0, z_0), we have three linear differential equations for determining r_0, s_0 and t_0:

$$r_0 H_0 + 2s_0 K_0 + t_0 L_0 = -M_0 \tag{B.51}$$

$$r_0\left(\frac{dx}{d\sigma}\right)_0 + s_0\left(\frac{dy}{d\sigma}\right)_0 = \left(\frac{dp}{d\sigma}\right)_0 \tag{B.52}$$

$$s_0\left(\frac{dx}{d\sigma}\right)_0 + t_0\left(\frac{dy}{d\sigma}\right)_0 = \left(\frac{dq}{d\sigma}\right)_0 \tag{B.53}$$

The determinant that is linked to the unknown quantities r_0, s_0, and t_0 is generally different from zero. It is termed Δ, and we have:

$$\Delta = \begin{vmatrix} H_0 & 2K_0 & L_0 \\ \left(\frac{dx}{d\sigma}\right)_0 & \left(\frac{dy}{d\sigma}\right)_0 & 0 \\ 0 & \left(\frac{dx}{d\sigma}\right)_0 & \left(\frac{dy}{d\sigma}\right)_0 \end{vmatrix} \tag{B.54}$$

$$= H_0\left(\frac{dy}{d\sigma}\right)_0^2 - 2K_0\left(\frac{dx}{d\sigma}\right)\left(\frac{dy}{d\sigma}\right) + L_0\left(\frac{dx}{d\sigma}\right)^2 \qquad (B.55)$$

Also different from zero are the determinants δ_1, δ_2, and δ_3, such as

$$\delta_1 = \begin{vmatrix} H_0 & M_0 & L_0 \\ \left(\dfrac{dx}{d\sigma}\right)_0 & -\left(\dfrac{dp}{d\sigma}\right)_0 & 0 \\ 0 & -\left(\dfrac{dq}{d\sigma}\right)_0 & \left(\dfrac{dy}{\alpha\sigma}\right)_0 \end{vmatrix} \qquad (B.56)$$

At the point (x_0, y_0, z_0) of $\mathscr{L}$ we obtain

$$r_0 = \frac{\delta_1}{\Delta} \qquad s_0 = \frac{\delta_2}{\Delta} \qquad t_0 = \frac{\delta_3}{\Delta} \qquad (B.57)$$

It can be shown that the derivatives of higher orders can be determined if r_0, s_0, and t_0 are known. Successive determinants, which come into play to determine the problem in its entirety through the Taylor series, are closely connected with the first ones. The determinant of the unknown quantities that intervene to determine the third derivatives is Δ^1, such as $\Delta^1 = \Delta^2$. The three determinants δ_1, δ_2, and δ_3 are shown to be linearly interdependent; δ_2 is preferentially considered.

All the curves $\mathscr{L}$, along which the Cauchy problem is undetermined, are termed *characteristics* and are denoted by C. Along these curves, $\Delta = 0$, $\delta_2 = 0$, and, of course,

$$dz = d\Phi = p\, dx + q\, dy \qquad (B.58)$$

The first condition ($\Delta = 0$) leads to

$$H\, dy^2 - 2K\, dx\, dy + L\, dx^2 = 0 \qquad (B.59)$$

and the second condition ($\delta_2 = 0$) requires

$$H\, dp\, dy + M\, dx\, dy + L\, dx\, dq = 0 \qquad (B.60)$$

The first equation of this group can be rewritten as

$$H\left[\frac{dy}{dx}\right]^2 - 2K\frac{dy}{dx} + L = 0 \qquad (B.61)$$

where dy/dx gives the slope of the curve c that represents the projection of the characteristic C in the xy plane. We set

$$\frac{dy}{dx} = \mu(x, y, \Phi, p, q) \qquad (B.62)$$

Table B.1

Characteristics of the first type I, μ_1

$$dy = \mu_1(x, y, \Phi, p, q)\, dx$$
$$H[x, y, \Phi, p, q]\,[dp + \mu_2\, dq] + M\,[\;]\, dx = 0$$
$$d\Phi = p\, dx + q\, dy$$

Characteristics of the second type II, μ_2

$$dy = \mu_2(x, y, \Phi, p, q)\, dx$$
$$H[x, y, \Phi, p, q]\,[dp + \mu_1\, dq] + M\,[\;]\, dx = 0$$
$$d\Phi = p\, dx + q\, dy$$

and find that

$$H\mu^2 - 2K\mu + L = 0 \qquad (\text{B.63})$$

which leads to two real or imaginary values of μ (μ_1 and μ_2) for a partial differential equation of second order.

The second equation can be written as

$$H\, dp\, \mu_1\, dx + L\, dq\, dx + M\, dx^2\, \mu_1 = 0 \qquad (\text{B.64})$$

Dividing by $\mu_1\, dx$ and taking into account the value of the product $\mu_1\mu_2$, we obtain

$$H\, dp + \frac{L}{\mu_1}\, dq + M\, dx = 0 \qquad (\text{B.65})$$

$$H(dp + \mu_2\, dq) + M\, dx = 0 \qquad (\text{B.66})$$

Finally, along a characteristic, the line functions x, y, Φ, p, and q are linked by three relations (Table B.1). We derive μ_1 and μ_2 from the equation

$$H\mu^2 - 2K\mu + L = 0 \qquad (\text{B.67})$$

These systems are called Pfaff forms. In the general case, they do not constitute a system of ordinary differential equations. In numerous cases, they degenerate into a set of ordinary differential equations, which leads to the solution of the problem as shown later.

The existence or nonexistence of some characteristic curves depends on the equation

$$H\mu^2 - 2K\mu + L = 0$$

If the two roots μ_1 and μ_2 are imaginary, the characteristics do not exist, and the equation is of the elliptic type. If the roots are real, the characteristics do exist, and the equation is of the hyperbolic type. If the roots are equal, the equation is of the parabolic type.

B.5   PROPERTIES OF THE CHARACTERISTICS

The importance of characteristics in the theory of partial differential equations is due to their following properties:

a. If our integral surface is associated with a partial differential equation of the second order, two series of characteristics can be drawn on this surface. On this surface, z or Φ, p and q are functions of x and y; therefore, we can express H, L, and M as functions of x and y. Let us suppose x and y to be constrained to verify the equation

$$H\left(\frac{dy}{dx}\right)^2 - 2K\frac{dy}{dx} + L = 0 \qquad (B.68)$$

which turns into an ordinary differential equation whose coefficients depend on x and y only. Two curves can generally be drawn in the xy plane, each of which corresponds to a curve drawn on the surface. We can show that such curves are characteristics.

For each curve, the previous condition is verified and also the connection between dx, dy, $d\Phi$, p, and q:

$$d\Phi = p\,dx + q\,dy \qquad (B.69)$$

because elementary displacement is made on the surface. As for the third relation,

$$H[dp + \mu_1\,dq] + M\,dx = 0 \qquad (B.70)$$

it must also be verified because along $\mathcal{L}$ the Cauchy problem has no solution if $\Delta = 0$; that is,

$$H\left(\frac{dy}{dx}\right)^2 - 2K\frac{dy}{dx} + L = 0 \qquad (B.71)$$

and $\delta_2 \neq 0$. Accordingly,

$$H[d\rho + \mu_1\,dq] + M\,dx = 0 \qquad (B.72)$$

We have correctly assumed that the Cauchy problem has a solution that is associated with the existence of an integral surface.

The curve drawn on the surface obeys the required equations (Eqs. B.68, B.69, B.71) to be a characteristic curve. Accordingly, two characteristics C_1 and C_2 belonging to two families can be drawn from each point of the surface integral. The generation of a surface integral by a characteristic is a straightforward consequence of this theorem. The two curves defined by Eq. B.68 are, of course, the projections c_1 and c_2 on the xy plane of the two characteristics C_1 and C_2 (see Fig. B.4).

b. If two integral surfaces are tangent to each other along the curve (see Fig. B.5), the curve is a characteristic; otherwise, the Cauchy problem would admit a unique solution.

The mechanical role of the characteristics becomes evident: Two integral surfaces can be matched along characteristics only. One of these characteristics can represent the undisturbed medium, whereas

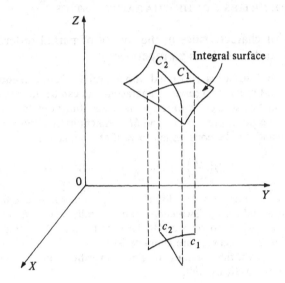

Fig. B.4   Characteristics and integral surface

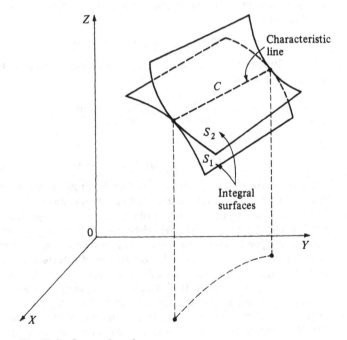

Fig. B.5   Integral surfaces

the second represents the slightly disturbed medium. Therefore, characteristics are associated with wavefronts. The determination of the wave fronts turns into the determination of the characteristics.

It is easy to understand from these remarks that characteristics are real if the partial differential equation is hyperbolic. The d'alembertian operator is the simplest operator associated with hyperbolic equations:

$$\frac{\partial^2}{\partial t^2} - a^2 \frac{\partial^2}{\partial x_i^2} \tag{B.73}$$

The laplacian operator

$$\frac{\partial^2}{\partial x^2} + \frac{\partial^2}{\partial y^2} + \frac{\partial^2}{\partial z^2} \tag{B.74}$$

is the simplest operator associated with elliptic equations, and the diffusion operator

$$\frac{\partial}{\partial t} - \frac{\partial^2}{\partial x_i^2}$$

is a parabolic operator.

When characteristics exist, they can be presented in several systems of coordinates. We previously saw the equations of the characteristics defined in the xy plane. Another representation could be given using the pq plane. The velocities being associated with the first derivatives of Φ ($u = p = \partial\Phi/\partial x$ and $v = q = \partial\Phi/\partial y$), the pq plane is the hodographic plane.

B.6   REMARKS

The theory of characteristics is often presented in a more restrictive way. The coefficients of the partial differential equations are assumed to be dependent on x and y only, so the characteristics are determined by the unique equation

$$H\left[\frac{dy}{dx}\right]^2 + 2K \frac{dy}{dx} + L = 0 \tag{B.75}$$

The roots of this equation can be written as

$$\frac{dy}{dx} = \frac{K \pm \sqrt{K^2 - LH}}{L} = \begin{cases} -n \\ -m \end{cases} \tag{B.76}$$

where m and n are two constants if H, K, and L are also constants; otherwise, they are functions of x and y. That leads to

$$x + my = C_1 = X(x, y) \tag{B.77}$$

$$x + ny = C_2 = Y(x, y) \tag{B.78}$$

In this new system of coordinates, a hyperbolic equation becomes

$$2 \frac{\partial^2 \Phi}{\partial X \, \partial Y} + M(X, Y) = 0 \tag{B.79}$$

which is its canonical form.

Elliptic equations can also be reduced to a canonical form by using a new system of coordinates:

$$X = \xi + i\eta \qquad Y = \xi - i\eta$$

which leads to

$$\frac{\partial^2 \Phi}{\partial \xi^2} + \frac{\partial^2 \Phi}{\partial \eta^2} = F(\xi, \eta) \tag{B.80}$$

For a parabolic equation, we find that

$$\frac{\partial^2 \Phi}{\partial Y^2} = F(X, Y) \tag{B.81}$$

If the three quantities H, K, and L are not constants, the equation is not necessarily of a unique type in the entire plane. The Tricomi equation

$$\frac{\partial^2 \Phi}{\partial x^2} + x \frac{\partial^2 \Phi}{\partial y^2} = 0 \tag{B.82}$$

changes type according to the sign of x.

B.7   SYSTEMS OF PARTIAL DIFFERENTIAL EQUATIONS

The system is assumed to refer to a two-dimensional flow so that two independent variables do exist. The system consists of n partial differential equations for n unknown functions $u_i(x, y)$. It can be written in the form

$$\Lambda_i(u) = \sum_{j=1}^{n} \left(a_{ij} \frac{\partial u_j}{\partial x} + b_{ij} \frac{\partial u_j}{\partial y} \right) + d_i = 0 \tag{B.83}$$

This set of equations can be presented in the matrix form

$$A \frac{\partial u}{\partial x} + B \frac{\partial u}{\partial y} + D = 0 \tag{B.84}$$

where A is the matrix $[a_{ij}]$, $B = [b_{ij}]$, $D = D_i$. The solution can be formulated in a vectorial form in which the u_j components of the function u are to be determined.

We begin to investigate whether a linear transformation

$$L[l_{ij}(x,\ y,\ u)]$$

does exist. The following properties are required for L:

$$\begin{cases} \det L \neq 0 \\ LA = CLB \end{cases} \tag{B.85}$$

in which C is a diagonal matrix. Such a connection between A and B is not a priori. We must determine what conditions are needed to obtain $LA = CLB$. If this relation is introduced, we have

$$LA\,\frac{\partial u}{\partial x} + LB\,\frac{\partial u}{\partial y} + LD = 0 \tag{B.86}$$

We set $LB = A^* = [a_{ij}^*]$ and $LD = D^* = [d_i^*]$, which gives

$$CLB\,\frac{\partial u}{\partial x} + LB\,\frac{\partial u}{\partial y} + LD = 0 \tag{B.87}$$

$$CA^*\,\frac{\partial u}{\partial x} + A^*\,\frac{\partial u}{\partial y} + D^* = 0 \tag{B.88}$$

The ith equation can be written as

$$\sum_{j=1}^{n} a_{ij}^*\left(\frac{\partial u_j}{\partial y} + c_{ii}\,\frac{\partial u_j}{\partial x}\right) + d_i^* = 0 \tag{B.89}$$

The components ξ, η of a unit vector such as $c_{ii} = \xi/\eta$ being introduced, we find that

$$\frac{\partial u_j}{\partial y} + c_{ii}\,\frac{\partial u_j}{\partial x} = \left(\eta\,\frac{\partial u_j}{\partial y} + \xi\,\frac{\partial u_j}{\partial x}\right)\frac{1}{\eta} \tag{B.90}$$

This last quantity is proportional to the derivative of u in the ξ, η direction. In other words, it is the slope of the integral surface in this direction. Accordingly, the line, whose direction cosines are given by Eq. B.90, is drawn on the surface; and it can be used to generate the integral surface. Simultaneously, the characteristics of the system are defined.

Let us determine the coordinates imposed by Eq. B.85.

$$\sum_{k=1}^{n} l_{ik}(a_{kj}) = \sum_{k=1}^{n} c_{ii}l_{ik}b_{jk} \qquad j = 1, 2, \ldots, n \tag{B.91}$$

$$\sum_{k=1}^{n} l_{ik}(a_{kj} - c_{ii}b_{kj}) \qquad j = 1, 2, \ldots, n \tag{B.92}$$

That requires $\det(A - c_{ii}B)$ to be zero, which leads to n real or imaginary values of c_{ii} for a system consisting of n unknown functions. The relation given by Eq. B.89 is called the *canonical form* of the system.

The method operates for a quasilinear system, which means that a_{ij}, b_{ij}, d_i and consequently c_{ii} are functions of the $(n + 2)$ variables (x, y, u_i). We do not discuss the required conditions of a_{ij}, b_{ij}, and d_i. If all these functions possess derivatives of any order, the method is successful.

Examples:

$$\frac{\partial^2 \Phi}{\partial t^2} = c^2 \frac{\partial^2 \Phi}{\partial x^2} \qquad \Phi\,(x,\,0) = f_1(x)$$

$$\frac{\partial \Phi}{\partial t}\,(x,\,0) = f_2(x) \tag{B.93}$$

This equation is equivalent to the system

$$\frac{\partial y}{\partial t} = c\,\frac{\partial \omega}{\partial x} \qquad y(x,\,0) = f_1(x)$$

$$\frac{\partial \omega}{\partial t} = c\,\frac{\partial y}{\partial x} \qquad \omega(x,\,0) = \frac{1}{c}\int_0^x f_2(x) \tag{B.94}$$

The corresponding matrix form can be written

$$\begin{bmatrix} 1 & 0 \\ 0 & 1 \end{bmatrix}\begin{bmatrix} \partial y/\partial t \\ \partial \omega/\partial t \end{bmatrix} + \begin{bmatrix} 0 & -c \\ -c & 0 \end{bmatrix}\begin{bmatrix} \partial y/\partial x \\ \partial \omega/\partial x \end{bmatrix} = \begin{bmatrix} 0 \\ 0 \end{bmatrix} \tag{B.95}$$

which leads to

$$|A - \lambda B| = \begin{vmatrix} 1 & \lambda c \\ \lambda c & 1 \end{vmatrix} \Rightarrow 1 - \lambda^2 c^2 = 0$$

$$\lambda = \pm\frac{1}{c} \tag{B.96}$$

In this hyperbolic system, the wave speed is equal to c.

C   Perturbation Theory

Perturbation theory is a collection of methods for the systematic analysis of the global behavior of solutions to differential and difference equations. The general procedure in perturbation theory is to identify a small parameter, usually denoted ε, such that when $\varepsilon = 0$ the problem becomes soluble.

A perturbative solution is constructed by local analysis about $\varepsilon = 0$ as a series of powers of

$$y(x) = y_0(x) + \varepsilon y_1(x) + \varepsilon^2 y_2(x) \tag{C.1}$$

This series is called a *perturbation series*. The attractive feature of this method is to give $y_n(x)$ in terms of $y_0(x) \cdots y_{n-1}(x)$ as long as the problem obtained by setting $\varepsilon = 0$ is soluble. The perturbation series for $y(x)$ is, of course, local in ε but is global (now local) in x. If ε is very small, we can expect $y(x)$ to be conveniently approximated by only a few terms of the perturbation series.

It happens that perturbation series diverge for all $\varepsilon \neq 0$. ε is not necessarily a small parameter, and so the optimal asymptotic approximation may give poor numerical results. Thus, to extract maximal information from perturbation theory it may be necessary to develop sophisticated techniques to accelerate the convergence of slowly diverging series. This refinement can be found in special texts devoted to this kind of problem. Here we give some elementary aspects of this problem as a necessary complement to the previous chapters.

C.1.a *Perturbation methods*

We could introduce boundary layer theory disregarding the general aspects of perturbations methods. In spite of the difficulties inherent in handling power series that may be poorly convergent, it is interesting to examine the possibilities of these methods in clarifying boundary layer theory as it applies to fluid mechanics.

In spite of their shortcomings, methods of perturbation theory can occasionally be so powerful that it is advisable to introduce a parameter ε temporarily as a factor into a difficult problem that initially has no small parameter and then later setting $\varepsilon = 1$ to recover the original problem. This parameter can be introduced in the equation itself or in the boundary conditions.

Whatever the specific aspects of the initial problem, the approach of perturbation theory is to decompose a difficult problem into an infinite number of relatively easy ones. Using perturbation methods, we expect, of course, that the first few steps will reveal the most important features of the solution and that the remaining steps will give small corrections only.

C.1.b *Perturbation analysis*

To present perturbation analysis, we begin with two examples: the first devoted to simple algebraic computation and the second to differential equations. Whatever the case under consideration, these three steps are fairly well defined:

1. Convert the original problem into a perturbation problem by introducing the small parameter ε.
2. Assume an expression for the answer in the form of a perturbation series and compute the coefficient of that series.
3. Recover the answer to the original problem by summing the perturbation series for the appropriate value of ε.

The introduction of a small parameter is flexible, and the first step may appear to be a delicate choice. Generally, it is preferable to introduce ε in such a way that the zeroth-order solution (which is the leading term in the perturbation series) is obtainable as an analytic expression. Anyway, the zeroth-order problem should be reasonably simple in order to introduce straightforward solutions.

Let us approximate the roots of the following cubic polynomial:

$$x^3 - 4.001x + 0.002 = 0 \qquad (C.2)$$

This problem can be stated as a perturbation problem after minor arrangements. For the present case, the trick is almost obvious. Instead of Eq. C.2, we deal with the one-parameter family of the polynominal equation

$$x^3 - (4 + \varepsilon)x + 2\varepsilon = 0 \qquad (C.3)$$

(When $\varepsilon = 0.001$, the original equation is reproduced.) We can present the solution of Eq. C.3 in the form of a perturbation series in powers of ε:

$$x(\varepsilon) = \sum_{n=0}^{\infty} a_n \varepsilon^n \qquad (C.4)$$

To obtain the first term of this series, we set $\varepsilon = 0$ and solve

$$x^3 - 4x = 0 \qquad (\text{C.5})$$

which yields

$$x(0) = a_0 = -2, 0, +2$$

A second-order perturbation approximation consists of writing

$$x_1 = -2 + a_1\varepsilon + a_2\varepsilon^2 + 0(\varepsilon^3) \qquad (\text{C.6})$$

Substituting this expression into the one-parameter equation results in

$$(-8 + 8) + (12a_1 - 4a_1 + 2 + 2)\varepsilon$$
$$+ (12a_2 - a_1 - 6a_1^2 - 4a_2)\varepsilon^2 = 0(\varepsilon^3) \qquad (\text{C.7})$$

ε being considered as a variable, we can conclude that the coefficient of each power of ε is separately equal to zero, which leads to

$$\varepsilon^1 \Rightarrow 8a_1 + 4 = 0 \qquad a_1 = -\frac{1}{2}$$

$$\varepsilon^2 \Rightarrow 8a_2 - a_1 - 6a_1^2 = 0 \qquad a_2 = \frac{1}{8}$$

Finally, we have

$$x_1 = -2 - \frac{1}{2}\varepsilon + \frac{1}{8}\varepsilon^2 + \cdots \qquad (\text{C.8})$$

The same procedure gives

$$x_2 = 0 + \frac{1}{2}\varepsilon - \frac{1}{8}\varepsilon^2 + 0(\varepsilon^3) \qquad x_3 = 2 + 0(\varepsilon) + 0(\varepsilon^2) + 0(\varepsilon^3)$$

As for $x_3 = 2$, it is the exact solution for all ε. These series converge for $\varepsilon < 1$.

Another example emphasizes the possibilities of the method. Let us consider the initial value problem

$$y'' = f(x)y \qquad y(0) = 1 \qquad y'(0) = 1$$

where $f(x)$ is assumed to permit further integrations. The solution of third problem has an analytic form for very special choices for $f(x)$.

First, we introduce ε in such a way that the unperturbed problem is soluble:

$$y'' = \varepsilon f(x)y \qquad y(= 0) = 1 \qquad y'(0) = 1$$

We assume also a perturbation expression for $y(x)$ of the form

$$y(x) = \sum \varepsilon^n y_n(x) \tag{C.9}$$

where $y_0(0) = 1$, $y'(0) = 1$, $y_n(0) = 0$, and $y'_n(0) = 0$.

The zeroth-order solution that satisfies the initial conditions is

$$y_0 = 1 + x \tag{C.10}$$

The nth-order problem is obtained by substituting $y(x) = \sum \varepsilon^n y_n(x)$ into $y'' = \varepsilon f(x)y$ and setting the coefficient of ε^n equal to zero. That leads to an iterative formulation:

$$y''_n = y_{n-1} f(x) \qquad y_n(0) = y'_n(0) = 0 \tag{C.11}$$

Perturbation theory has replaced the initial differential equation with a sequence of inhomogeneous equations whose solutions can be rewritten as

$$y_n = \int_0^x dt \int_0^t f(s)y_{n-1}(s)\, ds \tag{C.12}$$

This relation gives a simple iterative procedure for calculating successive terms in the perturbation series. Readers may wonder whether such a method is fundamentally different from the series methods that are classically introduced. Recall that the latter methods assume that $f(x)$ has a convergent Taylor expansion about $x = 0$, of the form

$$f(x) = \sum A_n x^n \tag{C.13}$$

and the solution $y(x)$ can be represented in the form of the series solution

$$y(x) = \sum a_n x^n \tag{C.14}$$

When we start from a linear equation, the series solution is guaranteed to have a radius of convergence equal to that of the first series $f(x) = \sum A_n x^n$. This close connection between the two radii of convergence does not appear in perturbation analysis. Moreover, perturbation analysis does not require $f(x)$ to be represented by a Taylor series expansion, though the integrals that appear in the iterative formulation must be meaningful.

Let us compare the two methods by applying them to the same differential equation (Fig. C.1):

$$y'' = -e^{-x}y \qquad y(0) = 1 \qquad y'(0) = 1 \tag{C.15}$$

The figure shows that the Taylor series are less powerful than the perturbation series.

Perturbation methods can be also used where boundary conditions

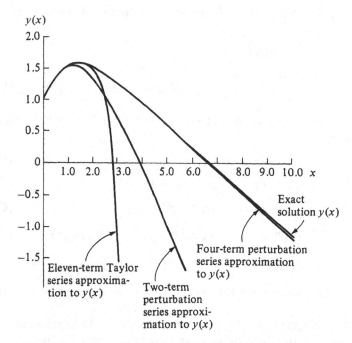

Fig. C.1 Comparison of Taylor series with approximations

are involved, presuming that the boundary conditions are not far from a simple form such as a circle.

Let us consider Laplace's equation for the interior of a region bounded by an "almost circle," in fact, by $r = R + \varepsilon \cos \theta$, where (r, θ) are polar coordinates. Here, $R > 0$ and $0 < \varepsilon \ll R$ are constants. In the interior of this region we want to find $\phi(r, \theta)$ such that

$$\Delta\phi = \frac{\partial^2\phi}{\partial r^2} + \frac{1}{r}\frac{\partial\phi}{\partial r} + \frac{1}{r^2}\frac{\partial^2\phi}{\partial\theta^2} = 0 \qquad (C.16)$$

with

$$\phi(R + \varepsilon \cos \theta, \theta) = f(\theta)$$

and where $f(\theta)$ is prescribed.

The solution of Laplace's equation is known when the boundary conditions are given on a circle. The perturbation procedure leads us to introduce the solution in the form of a series.

$$\phi(r, \theta) = \phi^{(0)}(r, \theta) + \varepsilon\phi^{(1)}(r, \theta) + \varepsilon^2\phi^2(r, \theta) + \cdots \qquad (C.17)$$

At this point, we write the boundary condition in the form of a Taylor series

$$\phi(R, \theta) + (\varepsilon \cos \theta)\phi_r(R, \theta)$$

$$+ \frac{1}{2} (\varepsilon \cos \theta)^2 \phi_{rr}(R, \theta) + \cdots = f(\theta) \qquad \text{(C.18)}$$

which leads to a sequence of problems for the underformed circular boundary:

$$\Delta\phi^{(i)} = 0 \qquad i = 0, 1, 2, \ldots \qquad \text{(C.19)}$$

$$\phi^0(R, \theta) = f(\theta) \qquad (\varepsilon = 0) \qquad \text{(C.20)}$$

$$\phi^1(R, \theta) = -\cos \theta \, \phi_2^{(0)}(R, \theta) \qquad \text{(C.21)}$$

$$\phi^2(R, \theta) = -\cos \theta \, \phi_{r(R, \theta)}^{(1)} - \tfrac{1}{2} \cos^2\theta \, \phi_{rr}^{(0)}(R, \theta) \qquad \text{(C.22)}$$

C.2 REGULAR AND SINGULAR PERTURBATION THEORY

The formal technique of perturbation theory can be considered a generalization of the ideas of local analysis of differential equations, so it is difficult to introduce regular or singular perturbation theory without refering to this kind of approach. The local approach to the solutions of an ordinary differential equation about an ordinary point $x = a$ was introduced by Cauchy using Taylor's power series. Near the point $x = a$, a series solution can be introduced about a. This power series can be used as parameter $x = a$, and the coefficients in the series can be computed recursively. The series is assumed to have a nonvanishing radius of convergence where linear ordinary differential equations are concerned. This radius is shown to be in close connection with the location of the singular points in the coefficients of the equations. For nonlinear equations, these points move in the representative plane, the locations depending on the initial conditions.

Recall that there are two diffusive types of series solutions to differential equations. A series solution about an ordinary point that leads to a Taylor series has a nonvanishing radius of convergence. A series solution about a singular point does not have this form except in some rare cases. For instance, we may expect an exceptional solution passing through the point $x = a$ and admitting a Taylor series representation. Instead, the series may either be a convergent series, not given in a Taylor series form, or it may be a divergent series. Series solutions about singular points often have the remarkable property of being meaningful near a singular point and yet not existing at the singular point. This situation is encountered in Laurent series, which is not meaningful at the point $x = a$ but is meaningful in an annular area. The discussion concerning the solution of ordinary differential equations is often carried out in the complex plane, which reinforces this analogy. In regular perturbation problems, the exact solution for small but nonzero values of $|\varepsilon|$ smoothly approaches the unperturbed or zeroth-order solution as ε goes to zero.

We define a singular perturbation problem as one whose perturbation series either does not take the form of a power series, or if it does, the power series has a vanishing radius of convergence. For some cases, no solution to the unperturbed problem exists. When a solution to the unperturbed problem does exist, its characteristic behavior is distinctly different from the behavior of those problems having an exact solution for arbitrarily small values of ε (zero value excepted). In other words, the exact solution for $\varepsilon = 0$ is fundamentally different from the neighboring solutions obtained in the limiting cases $\varepsilon \to 0$. This abrupt change in character is typical of any singular perturbation problem.

When dealing with a singular perturbation problem, we must take care to distinguish between the zeroth-order solution (the leading term in the perturbation series) and the solution of the unperturbed problem, because the latter may not even exist. To understand what the difficulty may be, it is convenient to work with equations whose complete solutions can be given in an analytical form.

Example: We deal with a second-order differential equation that presents some properties of a boundary layer. As in a boundary layer, the boundary conditions are not given at the same point. In fact, the introduction of the boundary conditions is rather questionable. Many problems in fluid mechanics should be presented as matching problems, some of them leading to eigenvalues problems. In a boundary layer, we have to match a condition at the edge of the boundary layer with another (a no-slip condition) at the wall. With jets, the problem is still more challenging because the two boundary conditions correspond to two zero values (velocity is null outside the jet). It is possible to show that such a situation leads to an eigenvalue problem. In fact, a constant of the motion is introduced through the momentum equations. The problem becomes more crucial with a wall jet with two zero values of velocity, at the wall (no-slip condition) and far outside the jet. The problem is not trivial because no evident constant of motion can be introduced (the friction forces at the wall reduce the fluid momentum). Another constant of motion is introduced. In that case, the problem is solved as an eigenvalue problem with two boundary conditions corresponding to $U = 0$. From a physical point of view, if the boundary conditions have zero values not complemented by a mechanical condition capable of fixing a given level (momentum, energy), an eigenvalue problem emerges for a regular solution, which enables us to fix the unknown level:

$$\varepsilon y'' - y' = 0 \qquad y(0) = 0 \qquad y(1) = 1 \qquad (C.23)$$

The associated unperturbed problem has no solution at all:

$$-y = 0 \qquad y(0) = 0 \qquad y(1) = 1$$

The solutions of this ordinary differential equation cannot have a regular perturbation expansion of the form

$$y = \sum_{n=0}^{\infty} y_n(x)\varepsilon^n \qquad (C.24)$$

because y_0 does not exist.

The highest derivative is multiplied by ε, and the limit $\varepsilon \to 0$, the unperturbed solution loses its ability to satisfy the two boundary conditions because a solution is lost.

In fact, the exact solution of the equation is easy to find:

$$y(x) = \frac{e^{x/\varepsilon} - 1}{e^{1/\varepsilon} - 1} \qquad (C.25)$$

The function is plotted in Fig. C.2 for several positive values of ε. For very small but nonzero ε, it is clear from the figure that it is almost constant except for a very narrow range near $x = 1$. This localized region of rapid variation is called the boundary layer. Thus, outside the boundary layer the exact solution satisfies the boundary condition $y(0) = 0$ and almost but not quite satisfies the unperturbed equations $y' = 0$.

Note that it is not obvious how to construct a perturbative approximation to a differential equation whose highest derivative is multiplied by ε with no solution available to the associated unperturbed problem.

C.3   INTRODUCTORY REMARKS

Perturbation theory appears to be far more general than boundary layer theory. In its general form it is extensively developed in specialized treatises. All these developments are interesting for an overall view of this kind of problem. For the sake of brevity, however, we restrict this presentation to cases closely related to the usual boundary layer theory encountered in fluid mechanics.

Extensions can be easily inferred from this case. In boundary layer theories the solution suffers a local breakdown as $\varepsilon \to 0$. A local break-

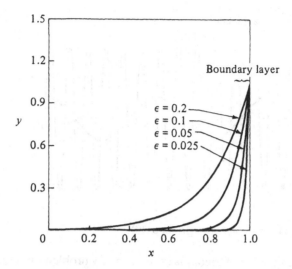

Fig. C.2   Influence of ε in the boundary layer

down occurs when the approximate solution is exponentially increasing or decreasing. This kind of behavior is called *dissipative* because it is associated with an ordinary differential equation having a strong positive or negative damping term. The equation

$$\varepsilon y'' + ay' + by = 0 \tag{C.26}$$

typically exhibits a dissipative behavior.

Some differential equations with small parameters have solutions that exhibit global breakdown. This is so for the boundary layer problem

$$\varepsilon y'' + y = 0 \qquad y(0) = 0 \qquad y(1) = 1 \tag{C.27}$$

which has the exact solution

$$y(x) = [\sin x \, \varepsilon^{-1/2}]/[\sin \varepsilon^{-1/2}] \tag{C.28}$$

This function rapidly becomes oscillatory for small values of ε and discontinuous when ε → 0 (Fig. C.3). The breakdown is global, occurring throughout the finite interval 0 ≤ x ≤ 1. This kind of behavior is called *dispersive*. A dispersive solution is wavelike with very small and slowly changing wavelengths and slowly varying amplitude of the x function. The wavelength of these oscillations is closely linked to the order of magnitude of the small parameter. Dissipative and dispersive phenomena are both characterized by exponential behavior associated with real or imaginary quantities. Multiscale analysis is a general collection of perturbation techniques that embodies the ideas of both boundary layer theory and WKB theory.

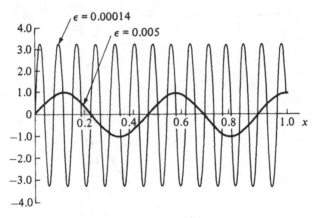

Fig. C.3 $y = (\sin x\varepsilon^{-1/2})/(\sin \varepsilon^{-1/2})$

For instance this sort of theory is useful in many problems of nonlinear oscillator equations with the appearance of secular terms in connection with the nonuniform validity of perturbation expansion for large values of t.

C.4   APPLICATION TO BOUNDARY LAYER THEORY

Among all the problems that can be introduced by singular perturbation theory, the boundary layer theory is the most important as far as viscous fluids are concerned. In this case we are still dealing with a differential equation whose highest derivative is multiplied by the perturbating parameter.

We define the boundary layer as a narrow region where the solution of a differential equation changes rapidly. By definition, the thickness of a boundary layer must approach 0 as $\varepsilon \to 0$. We shall limit our discussion to the cases of differential equations whose solutions exhibit only isolated, narrow regions of rapid variations; but we mentioned in the introduction that it is possible for a solution of a perturbation problem to undergo rapid variation over an extended region whose size does not approach zero with ε.

Our attention is focused on a simple boundary value problem whose solutions exhibit a boundary layer structure (see Fig. C.4):

$$\varepsilon y'' + (1 + \varepsilon)y' + y = 0 \tag{C.29}$$

The exact solution of this equation is

$$y(x) = \frac{(e^{-x} - e^{-x/\varepsilon})}{(e^{-1} - e^{-1/\varepsilon})} \qquad y(0) = 0 \qquad y(1) = 1 \tag{C.30}$$

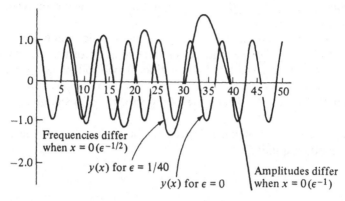

Fig. C.4 $y'' + (1 - \varepsilon x)y' + y = 0$

As ε approaches zero, this solution becomes discontinuous at $x = 0$. For very small values of ε, the solution slowly varies for $\varepsilon \ll x < 1$, whereas it undergoes an abrupt and rapid change for $0 \le x \le O(\varepsilon)$, where $O(\varepsilon)$ means that the thickness of the boundary layer is a function of ε as ε approaches zero. The region of slow variation of $y(x)$ is called the *outer region;* the region of rapid variations of $y(x)$ is called the *inner region* or *boundary layer region*.

We want to study the mathematical structure of the boundary layer inner and outer parts and the intermediate limit. This approach should enable us to loosely formalize and restate concepts that appear to be sufficient in many technical cases. In our boundary layer chapter, an attempt was made to match the outer (potential flow) and the inner solution (boundary layer). In this form, the problem is somewhat vague and refinements are required in many cases. The ability to solve the Navier–Stokes equations with big computers makes this sort of refinement less useful. However, the concept of boundary layer is so important in fluid dynamic theory that it is worthwhile to get acquainted with the mathematical basis of this concept because it is so much less time consuming.

The linear differential equation

$$\varepsilon y'' + (1 + \varepsilon)y' + y = 0 \qquad y(0) = 0 \qquad y(1) = 1 \qquad \text{(C.31)}$$

has the exact solution mentioned earlier:

$$y(x) = \frac{(e^{-x} - e^{-x/\varepsilon})}{(e^{-1} - e^{-1/\varepsilon})} \qquad \text{(C.32)}$$

For $\varepsilon \to 0$, it behaves as a boundary layer.

In fact, the function $y(x)$ has two components: e^{-x} is a slowly varying function on the entire interval $[0, 1]$ and $e^{-x/\varepsilon}$ is a rapidly varying function in the boundary layer $0 < x < \delta$ when $\delta = 0(\varepsilon)$ is the thickness of the boundary layer.

In fluid mechanics, the boundary layer development depends independently on $1/Re \sim \varepsilon$ and (x, y). The solution of the last differential equation can be considered a function of two independent variables x and ε. The aim of the analysis is to find a global approximation to y as a function of x. This is achieved by performing a local analysis of y as $\varepsilon \to 0_+$.

The outer limit of the solution can be deduced from the truncated differential equation

$$y'_{out} + y_{out} = 0 \tag{C.33}$$

which corresponds to

$$y_{out} = e^{1-x} \tag{C.34}$$

It can also be obtained from the general solutions by choosing $\delta \ll x \leq 1$ and allowing $\varepsilon \to 0$.

Because the outer limit is a first-order differential equation, its solution cannot satisfy both boundary conditions $y(0) = 0$ and $y(1) = 1$. It satisfies the external condition $y(1) = 1$ only. This outer part is assumed by the first component of the solution. The difference between the outer limit of the exact solution and the exact solution itself is exponentially small in the limit $\varepsilon \to 0$ when $\delta \ll x$.

The unperturbed differential equation has no solution that satisfies the two boundary condition $y(0) = 0$ and $y(1) = 1$, and so we have to deal with a singular perturbation. In other words, the limit of the complete solution is not everywhere close to the solution of the unperturbed differential equation.

Let us consider the inner part of the solution. The exact solution satisfies the boundary condition $y(0) = 0$ in the neighborhood of $x = 0$ $[x \leq 0(\varepsilon)]$. The behavior of this part of the solution is illucidated by introducing the x variable, a sort of scaling corresponding to ε so that we write

$$x = \varepsilon X \tag{C.35}$$

where X is the inner variable (η in boundary layer theory). y varies rapidly as a function of x but slowly as a function X. It is precisely the role of this adapted scaling.

Using the new variable X, we can write the differential equation

$$\frac{1}{\varepsilon} \frac{d^2Y}{dX^2} + \left[\frac{1}{\varepsilon} + 1\right] \frac{dY}{dX} + Y = 0 \tag{C.36}$$

with

$$Y(X) = y(x) \quad \text{and} \quad \varepsilon \frac{dy}{dx} = \frac{dY}{dX} \quad \varepsilon^2 \frac{d^2y}{dx^2} = \frac{d^2Y}{dX^2} \qquad \text{(C.37)}$$

When ε approaches zero, we get

$$\frac{d^2 Y_{in}(X)}{dX^2} + \frac{dY_{in}(X)}{dX} = 0 \qquad \text{(C.38)}$$

This equation admits a solution

$$Y(X) = e(1 - e^{-x}) \qquad \text{(C.39)}$$

which satisfies the boundary $Y_n(0) = 0$

We can extract from the general solution

$$y(x) = Y(X) = \frac{e^{-\varepsilon X} - e^{-X}}{e^{-1} - e^{-1/\varepsilon}} \underset{\varepsilon \to 0}{\Rightarrow} e(1 - e^{-X}) \qquad \text{(C.40)}$$

Boundary layer analysis is extremely useful because it enables us to construct an approximate solution of a differential equation even though an exact answer is not attainable. This is because the inner and the outer limits of an insoluble differential equation are often soluble.

In the last example, the complete differential equation has been chosen tractable just to make all the steps of the computations understandable. After y_{in} and y_{out} have been determined, they must be asymptotically matched. This matching assumes the existence of an overlapping region. In the case under consideration, such as situation does exist:

$$y_{out} = e^{1-x} \underset{x \to 0}{\Rightarrow} e \qquad y_{in} = e - e^{1-X} \underset{X \to \infty}{\Rightarrow} e \qquad \text{(C.41)}$$

In most cases a simple common limit for the outer and inner solution is not attainable. This limit is generally not independent of x and X. In the case under consideration, that provides a second boundary condition for the inner equation

$$\frac{d^2 Y_{in}(x)}{dX^2} + \frac{dY_{in}(X)}{dX} = 0 \qquad \text{(C.42)}$$

$$Y_{in}(\infty) = e \qquad \text{(C.43)}$$

Occasional simplicity should not dissimulate the central difficulty of the problem. Because the ability to construct a matched asymptotic expression depends on the existence of the overlapping region, how can we know whether the inner and outer limits of the differential equation have a common-region validity? To answer this question, we perform a complete perturbative solution of the equation

$$\varepsilon y'' + (1 + \varepsilon)y' + y = 0 \qquad y(0) = 0 \qquad y(1) = 1 \qquad (C.44)$$

to all orders in powers of ε.

First, we seek a perturbation expansion of the outer solution

$$y_{\text{out}}(x) \sim \sum_{n=0}^{\infty} y_n(x)\varepsilon^n \qquad (\varepsilon \to 0_+) \qquad (C.45)$$

The boundary condition $y(1) = 1$ is restored by setting $y_0(1) = 1$, $y_1(1) = 0$, and $y_n(1) = 0$. Note that y_{out}, in $y_{\text{out}} = e^{1-x}$, is the first term $y_0(x)$ in the perturbation expansion.

Substituting $\Sigma\, y_n(x)\varepsilon^n$ into the global differential equation and taking into account the powers of ε we obtain a sequence of differential equations:

$$y_0' + y_0 = 0 \qquad y_0(1) = 1$$

$$y_n' + y_n = -y_{n-1}'' - y_{n-1}' \qquad y_n(1) = 0 \qquad n \geq 1 \qquad (C.46)$$

The solutions to these equations are

$$y_0 = e^{1-x} \qquad y_n = 0 \qquad n \geq 1$$

Thus, the leading term of the outer solution $y_{\text{out}} = e^{-1-x}$ is correct to all orders in perturbation theory.

Next, let us perform a similar expansion of the inner solution. We write

$$Y_{\text{in}}(X) \sim \sum_{n=0}^{\infty} \varepsilon^n \, Y_n(X) \qquad \varepsilon \to 0 \qquad (C.47)$$

The boundary condition $y(0) = 0 \Rightarrow Y_{\text{in}}(0) = 0$ is restored by setting $Y_n(0) = 0$. A new sequence of differential equation is introduced:

$$Y_0'' + Y_0' = 0 \qquad Y_0(0) = 0 \qquad (C.48)$$

$$Y_n'' + Y_n = -Y_{n-1}'' - Y_{n-1} \qquad Y_n(0) = 0 \qquad n \geq 1 \qquad (C.49)$$

These equations can be solved by means of the integrating factor e^x, which leads to

$$Y_0(X) \doteq A_0(1 - e^{-X})$$

$$Y_n(X) = \int_0^X [A_n\, e^{-z} - Y_{n-1}(z)]\, dz$$

where the A_n are arbitrary integration constants.

The solution of the outer solution can be transformed into a power series of $x = \varepsilon X$, which yields

$$y_{\text{out}}(x) = e^{1-x} = e^{1-\varepsilon X} = e\left(1 - \varepsilon X + \frac{\varepsilon^2 X^2}{2!} - \frac{\varepsilon^3 X^3}{3!} + \cdots\right) \qquad (C.50)$$

For large values of X, the basic term of the inner solution yields

$$Y_0(X) = A_0 \qquad (C.51)$$

This basic term is to be compared with the first term of $y(x)$, which leads to $A_0 = e$. Now Y_0 is known, and we can compute Y:

$$Y_1(X) = (A_1 + A_0)(1 - e^{-x}) - eX$$

For large values of X, we find that

$$Y_1(X) = (A_1 + A_0) - eX$$

This term must be matched with the second term of the series available for $y_{\text{out}}(x)$, which yields

$$(A_1 + A_0) - eX = -eX \qquad (C.52)$$

$$A_1 = -A_0 = -e \qquad Y_1(X) = -eX$$

Similarly,

$$Y_n(X) = e\left[\frac{(-1)^n}{n!}\right]X^n \qquad (C.53)$$

Hence, the full inner expansion is

$$\begin{aligned} Y_{\text{in}}(X) &= e\sum_{n=0}^{\infty} \varepsilon^n \frac{(-1)^n X^n}{n!} - e^{1-X} \\ &= e^{1-\varepsilon x} - e^{1-X} = e^{1-x} - e^{1-X} \end{aligned} \qquad (C.54)$$

This inner expansion is, of course, valid not only for values of X inside the boundary layer but also for large values of X.

The initial differential equation is sufficiently simple that a infinite-order uniform approximation can be computed through the two variables x (the initial variable) and $x = \varepsilon X$ (the expanded variable). The expression $y = e^{1-x} - e^{1-X}$ does not exactly converge to $y(x)$ because n, the order of perturbation theory, tends to infinity. It is equal to the complete solution

$$y(x) = \frac{e^{-X} - e^{-x/\varepsilon}}{e^{-1} - e^{-1/\varepsilon}} \qquad (C.55)$$

if we neglect $e^{-1/\varepsilon}$, which approaches zero with ε.

In this boundary layer theory, the singular character of the perturbation is intrinsic to both the inner and the outer expansions. The outer

expansion is singular because there is an abrupt change in the order of
the differential equation when $\varepsilon = 0$. By contrast, the inner expansion
is a regular perturbation expansion for finite X; however, because
asymptotic matching takes place in the limit $X \to \infty$, the inner expansion
also becomes singular.

C.5 BOUNDARY LAYER OF THICKNESS DIFFERENT FROM ε

In the boundary layer theory developed in fluid mechanics, the inertial
forces and the viscous forces should be of the same order of magnitude.
From this point of view, the determination of δ requires in some cases
the introduction of the notion of a distinguished limit, which involves a
dominant balance argument. To illustrate this technique briefly, we start
from the equation

$$\varepsilon y'' + a(x)y' + b(x)y = 0 \qquad y(0) = A \qquad y(1) = B \qquad \text{(C.56)}$$

which has a boundary layer at $x = 0$ if $a(x) > 0$. In the inner region,
we let

$$y(x) = Y_{in}(X) \qquad X = \frac{x}{\delta}$$

So we have

$$\frac{dy}{dx} = \frac{1}{\delta} \frac{dY_{in}(X)}{dX} \qquad \qquad \text{(C.57)}$$

$$\frac{d^2y}{dx^2} = \frac{1}{\delta^2} \frac{d^2Y_{in}(X)}{d^2X^2} \qquad \qquad \text{(C.58)}$$

The differential equation takes the form

$$\frac{\varepsilon}{\delta^2} \frac{d^2Y_{in}}{dX^2} + \frac{a(\delta X)}{\delta} \frac{dY_{in}}{dX} + b(\delta X)Y_{in} = 0 \qquad \text{(C.59)}$$

from which three cases should be considered.

1. $\delta \ll \varepsilon$. The Eq. C.59 becomes

$$\frac{d^2Y_{in}}{dX^2} = 0 \qquad \text{and gives} \qquad Y_n(X) = A + cX$$

which satisfies the boundary condition $Y_{in}(0) = A$. This inner limit
does not match the outer solution because $\lim_{x \to \infty} Y_{in}(X) = \infty$ unless
$c = 0$, but y_{out} is not generally equal to A.

2. $\varepsilon \ll \delta$. The equation becomes $dY_{in}/dX = 0$, so $Y_{in} = A$. Again,
matching with the outer solution is not possible because $y_{out}(0) \neq A$.

3. $\varepsilon = \delta$ gives the leading-order equation

$$\frac{d^2 Y_{in}}{dX^2} + a(0) \frac{dY_{in}}{dX} = 0$$

The case $\delta = \varepsilon$ is called a *distinguished limit*. It involves a nontrivial balance between two or more terms of the initial equation.

This situation makes a matching process between inner and outer solution possible. It is possible to show that a relation between ε and δ may be required by examining, for instance, the equation

$$\varepsilon y'' + x^2 y' - y = 0 \qquad y(0) = y(1) = 1 \qquad (C.60)$$

In this case, the choice $\delta = \varepsilon^{1/2}$ is the distinguished limit.

Index

advection, 85
  in boundary layers, 193
Alfvén waves, 226
Arnold, 137

basic equations, 7
Beltrami–Gromeka flows, 47
Belousov–Zhabotinsky reaction, 147
Bénard, 145
Bergé, 148
Bernoulli equation, 46, 71
body forces, 19
Blasius solution, 185
boundary layer, 163
  approximation for, 167
  equations of, 170, 180
  kinematic, 101, 192
  scaling of, 90
  separation in, 175, 194
  thermal, 106, 194
  three-dimensional, 207
buffer layer, 105

Cauchy's theorem, 57
chaos, 121
chaotic advection, 121, 133
characteristics
  in a boundary layer, 173
  in partial differential equations, 289
chemical processes, 97
circulation, 54
Clausius inequality, 253
combustion, 198
complex potential, 137
continuity equation, 9, 66
continuum, 1, 242
control volume, 243
correlation
  of velocity fluctuations, 91
  at two points, 92
Couette flow, 143
Crocco's theorem, 46
cycles, 245

Dalembertian, 213
diffusion
  of a scalar, 97
  of energy, 86
  in boundary layers, 193
dissipation in boundary layers, 103
dissipative system, 124, 141
Deborah number, 197
density, 1
derivatives
  Jaumann, 40
  material, 38, 40

objective, 39, 275
deviatoric tensor, 73
dielectric field, 22
diffusion of turbulent energy, 86
diffusion flames, 202
dissipation of energy, 85
dynamical systems, 121

Ekman layer, 196
electric field, 22
electromagnetism, 228
energy equation, 265
enthalpy, 249
entropy, 255
  of a perfect gas, 258
equation of motion, 19
equation of state, 21, 247
Euler's equation, 20
eulerian coordinate, 47

Feigenbaum, 145
first law of thermodynamics, 19
flames
  diffusion, 202
  influence of mixture ratio, 207
  premixed, 202
  speed of, 205
forces
  surface, 13, 19
  volume, 13
Fourier transform, 91
friction velocity, 102
fractal, 128
  dimension of, 131

hamiltonian systems, 122, 141
Hausdorff–Besicovitch dimension, 131
Helmholtz's theorem, 55, 72
Hopf bifurcation, 142

inertial zone of spectrum, 95, 98
information dimension, 154
internal waves, 218
inviscid fluid, 16
irreversible process, 245
irrotational flow, 46, 72
isothermal process, 253

Jaumann derivative, 40
jet
  free, 112
  structure of, 203

KAM theorem, 137
kinematic boundary layers, 101, 192
kinetic energy of turbulence, 86
Kolmogoroff assumption, 93
Kolmogoroff length scale, 96

319

Lagrange theorem, 54
lagrangian
    coordinates, 47
    description, 1
Landau, 144
Langevin's equation, 36
Laplace equation, 46
law of the wall, 77, 104
Lewis number, 98
logistic map, 126
Lorenz attractor, 131
Lyaponov exponent, 126, 153

magnetohydrodynamic waves, 231
Manneville, 145
Margoulis number, 108
material derivatives, 38, 40, 264
Maxwell model, 37
memory effects, 23, 34, 84, 198
mixing zone, 189
momentum equation, 45
Moser, 137

Navier–Stokes equations, 25, 29, 78, 82, 121
Newhouse, 145
Newtonian fluids, 31, 39
normal shock waves, 261

objective derivatives, 39, 275
orbits, 152

Partial differential equations
    in boundary layers, 177
    in mathematical physics, 281
pathline, 3
perfect fluid, 8
perturbation theory, 301
phase velocity, 223
Poincaré section, 137
Poisson's equation, 68
polytropic process, 250
Pomeau, 145
Prandtl number, 98, 109
premixed flames, 202
pressure gradient
    in boundary layers, 187
    role of, 45
pressure waves, 224
production of turbulence, 86
    in boundary layers, 103
    in wall jets, 106
pendulum, 124

rate of strain
    pressure correlation, 89
    tensor, 36
Rayleigh number, 143
relaxation modulus, 37
reversible process, 245
Reynolds number, 79
Reynolds stress tensor, 83
rheological properties, 13
Rotta's term, 89
Ruelle, 145

scaling, 40
    of turbulence, 90
second lay of thermodynamics, 253, 257
sensitivity to initial conditions, 141

separation
    in boundary layers, 175
    in three-dimensional boundary layers, 208
separatrix, 124
shear flow, 87
shear rate, 32
shear stress, 32, 188
shock waves, 261
singular perturbations, 175, 306
small oscillations method, 77
sound waves, 212
spectral tensor, 91
spectrum
    of turbulent energy, 91
    of a scalar quantity, 97
steady flow, 4, 6, 46
strain rate, 49
strange attractors, 131
streakline, 3
streamline, 3
stress tensor, 14
structural units, 41
surface forces, 13, 19
system, 243
Swinney, 147

Takens, 145
Taylor vortices, 143
thermal conductivity, 23
thermal boundary layer, 194
thermodynamic coordinates, 244
thermodynamic pressure, 28
thermodynamics, 243
torus, 145
turbulence, 71, 77
    and chaos, 141
    production in boundary layers, 103

unsteady flow, 6

Van der Pol equation, 125
velocity field, 32
viscosity, 32
viscoelastic materials, 35
viscous diffusion, 86
viscous sublayer, 105
volume forces, 13, 45
vortex filament, 47
vortex lines, 54
vortex rings, 52
vortex tube, 47
vorticity
    equation for, 45
    field, 61
    flux, 51
    law of, 45

wake, 113
    law of, 105
wall
    jet, 106, 190
    law of, 77, 104
waves
    dispersion of, 223
    internal, 218
    propagation of, 211, 214
    scattering of, 233
    sound, 212
    speed of, 215